Green Communication in 4G Wireless Systems

RIVER PUBLISHERS SERIES IN COMMUNICATIONS

Consulting Series Editors

MARINA RUGGIERI
University of Roma "Tor Vergata"
Italy

HOMAYOUN NIKOOKAR
Delft University of Technology
The Netherlands

This series focuses on communications science and technology. This includes the theory and use of systems involving all terminals, computers, and information processors; wired and wireless networks; and network layouts, procontentsols, architectures, and implementations.

Furthermore, developments toward new market demands in systems, products, and technologies such as personal communications services, multimedia systems, enterprise networks, and optical communications systems.

- Wireless Communications
- Networks
- Security
- Antennas & Propagation
- Microwaves
- Software Defined Radio

For a list of other books in this series, visit
http://riverpublishers.com/river_publisher/series.php?msg=Communications

Green Communication in 4G Wireless Systems

Editors

Shahid Mumtaz
Instituto de Telecommunicações
Aveiro, Portugal

and

Jonathan Rodriguez
Instituto de Telecommunicações
Aveiro, Portugal

LONDON AND NEW YORK

Published 2013 by River Publishers
River Publishers
Alsbjergvej 10, 9260 Gistrup, Denmark
www.riverpublishers.com

Distributed exclusively by Routledge
4 Park Square, Milton Park, Abingdon, Oxon OX14 4RN
605 Third Avenue, New York, NY 10158

First published in paperback 2024

Green Communication in 4G Wireless Systems / by Shahid Mumtaz, Jonathan Rodriguez.

Routledge is an imprint of the Taylor & Francis Group, an informa business

Publisher's Note
The publisher has gone to great lengths to ensure the quality of this reprint but points out that some imperfections in the original copies may be apparent.

While every effort is made to provide dependable information, the publisher, authors, and editors cannot be held responsible for any errors or omissions.

ISBN: 978-87-92982-05-6 (hbk)
ISBN: 978-87-7004-512-4 (pbk)

Table of Contents

Preface   vii

1. Introduction   1
Shahid Mumtaz and Jonathan Rodriguez

2. Overview of Spectral- and Energy-Efficiency Trade-off in
OFDMA Wireless System   9
*Kazi Mohammed Saidul Huq, Shahid Mumtaz, Jonathan Rodriguez and
Rui L. Aguiar*

3. Context Based Node Discovery Mechanism for Energy Efficiency in
Wireless Networks   33
*Muhammad Alam, Michele Albano, Ayman Radwan and
Jonathan Rodriguez*

4. Throughput Fairness Analysis of Reservation Protocols for
WiMedia MAC   53
*Muhammad Alam, Michele Albano, Ayman Radwan and
Jonathan Rodriguez*

5. Resource Allocation and Energy Calculation in WPANs Based
on WiMedia MAC   69
*Muhammad Alam, Shahid Mumtaz, Christos Verikoukis and
Jonathan Rodriguez*

6. Clustering Techniques for Energy Efficient Wireless
Communications   89
*Victor Sucasas, Hugo Marques, Jonathan Rodriguez and
Rahim Tafazolli*

7. Network Coding for Wireless Networking   117
Riccardo Bassoli, Hugo Marques, Jonathan Rodriguez and
Rahim Tafazzoli

8. Secure and Energy Efficient Vertical Handovers   139
Hugo Marques, Joaquim Bastos, Jonathan Rodriguez and
Rahim Tafazolli

9. Link Layer Modelling for Energy Efficient Performance
Evaluation in Wireless Cellular Networks   177
Valdemar Monteiro, Shahid Mumtaz, Jonathan Rodriguez and
Christos Politis

10. System Level Evaluation Methodology for Energy Saving   203
Valdemar Monteiro, Shahid Mumtaz, Alberto Nascimento,
Jonathan Rodriguez and Christos Politis

11. Conclusions   259
Shahid Mumtaz and Jonathan Rodriguez

Author Index   265

Subject Index   267

About the Editors   269

Preface

Throughout the global community, wireless communications have had a profound socio-economic impact, enriching our daily lives with a plethora of services from media entertainment to more sensitive applications such as e-commerce. Looking towards the future, although voice and SMS are still major sources of revenue, mobile traffic will account for a large chunk of the internet highway. To cope with this increased demand, operators are required to invest more in the core infrastructure, and deploy more advanced technologies. In fact, already in todays market operators are deploying over 120,000 new Base Stations (BSs) on a yearly basis across the world. Moreover, the fast evolvement of mobile technologies, from the 3rd Generation (3G) supporting 384 kb/s downlink in 2001 to the Long Term Evolution (LTE) supporting 300 Mb/s downlink in 2010, further increases the cost of network operators. However, although the data throughput per user is rapidly growing, the revenue per Mb is dropping, so that employing advanced spectral efficiency technology alone appears unable to maintain a profitable business. Consequently, service providers – and infrastructure vendors – are increasing their focus on cost reductions. The increasing demand in data and voice services alone is not the only cause for concern: energy management and conservation is now at the forefront of the political agenda.

The vision of Europe 2020 is to become a smart, sustainable and inclusive economy, and as part of these priorities the EU have set forth the 20:20:20 targets where greenhouse gas emissions should be reduced by 20%, whilst energy from renewable sources efficiency should be increased by that respective amount. In fact, in today's energy conscious society, Information and Communication Technology (ICT) accounts for 2% of the global CO_2 emissions. To place this in perspective, a medium sized cellular network uses as much energy as 170,000 homes.

Therefore, new solutions are required whereby operators can accommodate this additional traffic volume whilst reducing their investment in new infrastructures and beyond that significantly reducing their energy bill. Moreover, the EU political agenda in unison with expected growth in mobile

vii

data has identified cost and energy per bit reduction as a stringent design requirement for wireless networks of the future: namely 4G and beyond.

The green revolution towards a more energy conscious wireless communication infrastructure provided the context for this book, however the pieces of the jigsaw were put in place by the editorial team and the 4TELL research group within the Instituto de Telecomunicações, who not only have vast experience in green communications, but who are currently a major player in green research at the international research arena leading two major European projects, namely: the FP7 ICT-C2POWER project, based on the notion of "intelligent beyond 4G mobiles" that exploits context aware systems and the cooperation paradigm for reducing energy consumption at the mobile terminal; and the EUREKA CELTIC Green-T project that addresses green Device-to-Device communications and lightweight security. Moreover, the green initiative is here to stay and not just "a brand" that simply becomes out of fashion and forgotten in the darkest corner of the closet. The current energy targets are ambitious, and new business models will appear with the onset of 5th Generation services whose appetite for energy will grow. Therefore, new faces on the block need to appear to handover the research baton in order to ensure that we cross the finishing line with green winners medal. 4TELL is committed towards this cause through our training vehicle and participation in the Marie Curie project "GREENET", that aims to train early stage researchers to become tomorrows' Olympians on the green track event.

The context is green communications in 4G wireless technologies, the inspiration emanates from our involvement in European research and local initiatives spanning over 10 years, but this book would still not be possible if not for all the dedicated members of the 4TELL research group and external institutions such as the University of Surrey (UK), Kingston University (UK), University of Madeira (Portugal), and Universitat de Barcelona (Spain) who have contributed with chapters towards this book. The editors would also like to acknowledge the FP7-PEOPLE-IAPP CODELANCE and FCT SMARTVISION that have also contributed with new ideas on applications of Network Coding for promoting energy efficiency in future emerging technologies beyond 4G. Finally, the editors would like to especially thank Cláudia Barbosa, and Valdemar Monteiro who have spent endless moments assisting with the review and compilation of this book.

Shahid Mumtaz and Jonathan Rodriguez

1

Introduction

Shahid Mumtaz and Jonathan Rodriguez

Instituto de Telecomunições, Aveiro, Portugal
e-mail: smumtaz@av.it.pt

Mobile data traffic is growing at such an astonishing rate that current 3G networks are expected to saturate even with the elasticity provided by advanced signal processing techniques for maximising spectral efficiency, such as cross-layer design and MIMO, among others. On the other hand, services increasingly become more sophisticated requiring virtual broadband connectivity: service at any time, on any device at any place. These two drivers provided the springboard for the blueprint of a new generation of mobile telephony that can match the expectations of the operators and the consumer market. This standard – now known as LTE (Long-Term Evolutionary RAN) – is increasingly positioning itself as the answer for 4th Generation Mobile Network, although it still poses many open questions that need to be answered if it is to be a winning candidate for the 4G revolution. The key question arises from the surprising statistic given in Figure 1.1. It can be seen that mobile operators are continuously under pressure to enhance their infrastructure to deliver mobile services in a cost-effective way, however, their investment in new infrastructures does not always pay off since the average revenue per connection continues to reduce over time.

In fact, already in today's market operators are deploying over 120,000 new Base Stations (BSs) on a yearly basis across the world. Moreover, the fast evolvement of mobile technologies, from the 3rd Generation (3G) supporting 384 kb/s downlink in 2001 to the Long Term Evolution (LTE) supporting 300 Mb/s downlink in 2010, further increases the cost of network operators. However, and although the data throughput per user is rapidly

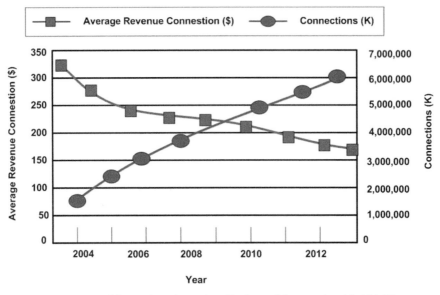

Figure 1.1 Mobile service price and traffic demand forecast (worldwide) [1].

growing, the revenue per Mb is dropping so that employing advanced spectral efficiency technology alone appears unable to help maintain a profitable business, requiring a new dimensioning metric in the network .

Another critical point for operators, which is gradually gathering concerns of the community at large, is the energy issue. It is reported that the total energy consumed by the infrastructure of cellular wireless networks, wired communication networks, and internet takes up more than 3% of the worldwide electric energy consumption nowadays [2] and that value is expected to increase rapidly in the future. A medium sized cellular network uses the same amount of energy as a total of 170,000 homes, with the cost of powering the existing BSs accounting for roughly 50% of a service provider's overall expenses. New solutions seem, therefore, required whereby operators can accommodate this additional traffic volume whilst decreasing their investment in new infrastructures and beyond that significantly reduce their energy bill. Figures 1.2(a) and 1.2(b) show a breakdown of power consumption in a typical cellular network and gives us an insight into the possible research avenues for reducing energy consumption in wireless communications.

Energy management and conservation is understandably now at the forefront of the political agenda. The vision of Europe 2020 is to become a smart,

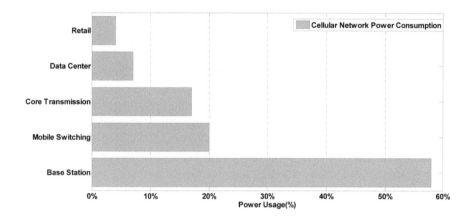

(a) Power consumption of a typical wireless cellular network [2]

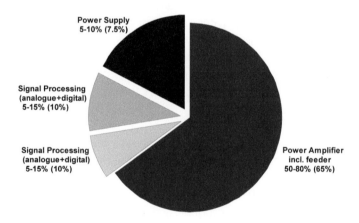

(b) Power consumption distribution in radio base stations [3]

Figure 1.2 Power consumption.

sustainable and inclusive economy, and as part of these priorities the EU have set forth the 20:20:20 targets where greenhouse gas emissions should be reduced by 20%, whilst energy from renewables, and energy efficiency should be increased by the same respective amount.

In today's energy conscious society, Information and Communication Technology (ICT) accounts for 2% of the global CO_2 emissions, a figure which is making the field of wireless communications (as an important part of ICT) invest in novel research paradigms for energy saving. Furthermore, mobile terminals in wireless systems necessitate energy saving since the development of battery technology is much slower when compared to the increase of energy consumption. Pursuing high energy efficiency (EE) will be, therefore, the trend for the design of future wireless communications. Moreover, the EU political agenda in unison with expected growth in mobile data has identified cost and energy per bit reduction as a stringent design requirement for mobile networks of the future.

During the past decades, much effort has been made to enhance network throughput. Different network deployments have been well investigated to improve area spectral efficiency, e.g., optimization of the number of base stations (BSs) in cellular networks and the placement of relay nodes in relay systems. Numerous resource allocation schemes have been proposed to assure quality-of-service (QoS) of each user and fairness among different users by exploiting multiuser diversity. Many advanced communication techniques, such as orthogonal frequency division multiple access (OFDMA), multiple-input multiple output (MIMO) techniques, and relay transmission, have been fully exploited in wireless networks to provide high spectral efficiency (SE). However, high network throughput usually implies large energy consumption, which is sometimes unaffordable for energy-aware networks or energy-limited devices.

Therefore the question towards how to reduce energy consumption while meeting throughput requirements in such networks and devices is now the new perspective towards network dimensioning that the research community is trying to answer. Several international research projects such as EARTH [3], C2POWER [4], and GREEN-T [5] among others, are formally targeting the first wave of energy efficient mobile networking architectures and techniques, and are pursuing solutions for different layers of the protocol stack and efficient network dimensioning.

Research results are highlighting that in network planning the impact of cell sizes affects the EE: reducing the cell size can increase the number of delivered information bits per unit energy for a given user density and total power in the given service area [6]. If sleep mode is introduced, the EE can be further enhanced. In addition, mixed cell deployment, e.g. using micro-cells at the edge of a macro-cell, is also an efficient way to save energy as well as to enhance the performance of cell edge users. At the Network layer, different energy efficient routing techniques have been developed to reduce energy consumption, and network coding is seen as incremental step forward to achieving link reliability and energy gain. For the medium access control (MAC) layer, protocols have been designed to efficiently utilize resources, such as power, time slots, and frequency bands, in order to reduce energy consumption. For the physical layer, different transmission techniques have been reconsidered from the EE point of view instead of traditional SE. Some cross-layer approaches have also been developed to obtain more gain over the independent layer design [7].

To ensure effective and inter-operable energy saving solutions, it is important to standardize fundamental mechanisms. For example 3GPP's TSG RAN WG3 is studying radio architecture aspects of energy saving for UMTS and LTE. The objective is to analyze energy saving potential and corresponding requirements for radio access networks. In addition, the first technologies for basic distributed energy saving functions are currently being developed.

3GPP's TSG SA WG5 is analyzing OAM aspects for energy saving in the TR 32.826 [7] study on Energy Saving Management by describing several use cases and corresponding requirements. The study shows what energy savings can be achieved in typical scenarios (e.g., with a day/night arrangement in urban networks) and what network management requirements result from approaches such as adapting power consumption to the actual load on the network elements. The OAM aspects of energy saving are of particular importance for coordinated energy saving schemes that can help to minimize operational cost while maintaining network performance for users at all times.

Energy efficiency in new generation networks is pivotal if operators are to maintain their competitive stance. There have been a plethora of new solutions towards achieving the delicate balance between maximising spectral efficiency and energy efficiency, some of which are being considered for standardization in future emerging standards such as LTE-A. The inspiration behind this book is energy efficiency in mobile networks, and we aim not only to summarize mature results on this dynamic field, but to go beyond state of the art and propose new paradigms that have the potential to go one step further to make green communications a reality. The organization of this book is as follows.

Chapter 2 discusses SE-EE trade-offs for OFDMA downlink systems in order to describe how combining adequate trade-offs allow us to make energy aware flexible radios and networks. Theoretical bounds for the single cell and multiple cell case are also presented. Finally, the effect of circuit power on these trends is analysed.

Chapter 3 discusses lightweight node discovery mechanisms for WPANs. It proposes new energy efficient discovery mechanisms that are essential for latching onto the benefits of cooperative mechanisms. A context based beaconing mechanism both for node discovery and cooperative cluster formation is presented, while the available beacon Information Elements (IEs) of UWB is exploited and a new Context Extraction Function (CxEF) presented. Beyond this, a detailed representation of bit mapping for energy and cooperation in beacons is also showcased. The chapter is concluded by simulation results, which demonstrate the potential energy saving that can be achieved using the proposed node discovery algorithm based on Context Awareness.

Chapter 4 discusses the analysis of the WiMedia reservation protocols. By means of simulations and analytical results, the chapter shows how the QoS provisioning issue is handled by the MAC protocols and how fairness is provided to users in terms of throughput and delay, shortening the time they spend trying to access the wireless medium. In turn, the net effect contributes to energy saving for data communication at the nodes, and adds to the Quality of Experience (QoE) of the user.

Chapter 5 discusses the solution for the resource allocation problem in WPANs based on the WiMedia MAC. WiMedia provides a decentralized MAC protocol for resource allocation, but still there are some issues related to the QoS provisioning which are highlighted in this chapter. The study presented suggests that UWB offers high data rates coupled with an energy consumption per bit which is significantly lower compared to other short range technologies.

Chapter 6 discusses the features and hardware capabilities of new generation mobile devices equipped with wireless interfaces such as Bluetooth and Wi-Fi, which have opened new doors for energy efficient wireless communication. Nevertheless, stability and scalability in ad hoc networks must be provided before taking advantage of energy saving in short range communication, and clustering algorithms are envisioned for that task. In this sense, among the state-of-the-art clustering algorithms, mobility-aware algorithms are considered the best solution for the dynamic nature of ad hoc networks in urban scenarios. Although there are many proposals for clustering algorithms in mobile networks, an adaptation is required for urban characteristics, where the mobility pattern and the environmental conditions have a big impact on the algorithm performance.

Chapter 7 discusses the theoretical framework for wireless network coding applications. Network coding research is increasingly gathering pace since it has been shown to be a powerful solution for energy efficient networks and video streaming applications. However, there are still open issues on how many of these solutions can be actually implemented in practice. This chapter investigates some of the most typical network coding solutions for implementation on real networks and highlights the challenges that lie ahead.

Chapter 8 discusses the secure and energy efficient vertical handover techniques. Such a handover can be triggered based on the maximization or minimization of specific parameters of the radio access, such as maximizing throughput or enhancing the energy efficiency of ongoing communications. Nonetheless vertical handovers raise security issues, even though there are security mechanisms for each access technology, a common security framework that enables authentication and key establishment from one technology to another is still missing. This chapter also includes a section with experimental results in an attempt to evaluate key aspects concerning energy efficiency in a VHO scenario modelled in a simulation platform based on the ns-2 network simulator, by simulating multiple VHOs under an IEEE 802.21 implementation. It is shown how the 802.21 functionality can be incorporated in ns-2 through add-on modules based in the 802.21 (draft 3) standard, developed by the National Institute of Standards and Technology (NIST).

Chapter 9 analyzes aspects related to link level with analysis on the energy consumption aspects. It gives a survey on the basics of simulation methodologies for wireless/cellular networks, including simulation models for traffic, mobility, channel characterization and the Link Level Interface (LLI) with specific emphasis on the energy efficiency analysis.

Chapter 10 extends the analysis given in Chapter 9 to understand how the link level performance interacts with the system level to form part of the system level model for analyzing system performance in cellular networks. Moreover, this chapter focuses on the dynamic resource allocation entity which is responsible for managing radio resources in wireless systems. The authors have identified the energy saving performance metrics and provide a holistic system level performance analysis using traditional scheduling policies from an energy saving perspective. This reveals that that scheduling policy plays an important role in reducing the energy consumption, highlighting that there is still much research to undertake on this area.

Chapter 11 presents the overall conclusion of the book.

References

[1] J. ca Ekholm, Forecast: Mobile services, worldwide, 2004–2013, 4q09 update, December 2009. [Online] available from http://www.gartner.com/id=1241715.

[2] G. Fettweis and E. Zimmermann, ICT energy consumption – Trends and challenges, in *Communications*, ser. 1. WPMC, 2008, no. WPMC 2008, pp. 2006–2009.

[3] M. Gruber, O. Blume, D. Ferling, D. Zeller, M. A. Imran, and E. C. Strinati, EARTH – Energy aware radio and network technologies, in *Proceedings of 2009 IEEE 20th International Symposium on Personal, Indoor and Mobile Radio Communications*. IEEE, pp. 1–5, September 2009.

[4] C2Power Project, http://www.ict-c2power.eu/.

[5] Green-T Project, http://greent.av.it.pt/.

[6] F. Meshkati, H. V. Poor, and S. C. Schwartz, Energy-efficient resource allocation in wireless networks, *IEEE Signal Processing Magazine*, vol. 24, no. 3, pp. 58–68, May 2007.

[7] S. Verdu, Spectral efficiency in the wideband regime, *IEEE Transactions on Information Theory*, vol. 48, no. 6, pp. 1319–1343, June 2002.

2

Overview of Spectral- and Energy-Efficiency Trade-off in OFDMA Wireless Systems

Kazi Mohammed Saidul Huq[1,2], Shahid Mumtaz[1],
Jonathan Rodriguez[1], and Rui L Aguiar[1,2]

[1]*Instituto de Telecomunições, Aveiro, Portugal*
[2]*Universidade de Aveiro, Aveiro, Portugal*
e-mail: kazi.saidul@av.it.pt

Abstract

Nowadays, there is a strong pressure to address the global warming problem at multiple levels. Wireless communication research is no exception: decreasing the carbon footprint of mass information communication technology (ICT) products and services is one of the major current research trends in telecommunication. Presently, 3% [1] of the world-wide energy is consumed by ICT infrastructures, inducing about 2% [1] of the world-wide CO_2 emissions. As conveyed data-volume increases very fast and wireless communications are increasingly used, network designs have pragmatically been refocused on energy-efficient designs able to limit CO_2 emissions. This approach is currently addressed as "Green Communications" in 4G networks. Significant energy savings in mobile networks can be expected by defining and standardizing energy-efficiency techniques, which include finding the trade-off between different engineering parameters such as spectral-efficiency (SE) and energy-efficiency (EE). This chapter discusses SE-EE trade-offs for OFDMA downlink systems in order to describe how combining adequate trade-offs allow us to make energy aware flexible radios and networks. Theoretical bounds for the single cell and multiple cell case are also presented. Finally, we analyzed the effect of circuit power on these trends.

Shahid Mumtaz and Jonathan Rodriguez (Eds.), Green Communication in 4G Wireless Systems, 9–32.

Keywords: 4G, spectral-efficiency, energy-efficiency, green.

2.1 Introduction

Next-generation wireless networks are anticipated to furnish increasingly high-speed Internet access anywhere and anytime. The popularity of mobile wireless devices such as smartphones, tablet PCs, etc. is inevitably accelerating the process with increased quality of service (QoS) demands, such as those presented by mobile video and gaming. The exponentially growing data traffic and the requirement for pervasive access have pushed a sudden expansion of network infrastructures and led to a rapid increase of energy demands. Therefore, it is becoming of major importance to mobile operators to maintain sustainable capacity growth and, at the same time, limit the power usage.

The increase of energy consumption in wireless networks directly results in increased greenhouse gas (GHG) emission, which has been considered a major threat to climate change and sustainable development. Yan Chen et al. [2] stated that "The European Union has acted as a leader in energy saving across the world and targeted a 20% GHG reduction. Other big countries such as China also promised to reduce the energy per unit of gross domestic product (GDP) by 20% and major pollution by 10 percent by 2020". The ever increasing pressure of social and environmental responsibility serves as another strong driving force for wireless operators to dramatically reduce energy consumption and carbon footprint. Subsequent initiatives have been already taken worldwide. For example, the Vodafone Group has declared to cut down its carbon emissions by 50% from its 2006–2007 existing of 1.23 million tons by 2020.[1]

To still meet the tasks raised by the high demands of data traffic and energy usage, this green evolution has turned into an immediate need for wireless networks today. The Third Generation Partnership Project (3GPP) community has foreseen energy-efficiency (EE) improvement as one of the most important areas that demand innovation for wireless standards beyond Long Term Evolution (LTE) [3].

As a consequence, with energy saving and environmental conservation becoming global demands and unavoidable trends, wireless communications researchers and engineers are changing their efforts towards EE oriented design, that is, green radio communications (GRC) [2]. Previously, several

---

[1] This information is from the research article [2].

research efforts already aimed at energy savings in wireless networks, such as designing ultra-efficient power amplifiers, reducing feeder losses, and introducing passive cooling [4]. Since these research efforts were kept separated, a global vision of what could be cumulatively achieved on energy saving, in a time-frame of 10–15 years, could not be formed [2]. Hence, innovative solutions that cannot be achieved via isolated efforts, using top-down architectures and joint design across all system levels and protocol stacks are now targeted by the research community [2], keeping EE as a central focus. While conventional wireless networks mainly emphasizes aspects of ubiquitous access and large system capacity, now research in wireless networks is mainly focusing on system capacity and spectral-efficiency (SE).

SE, defined as the system throughput per unit of bandwidth, is a widely adopted indicator for wireless network performance evaluation. The peak value of SE is always among the key performance indicators of 3GPP evolution, for example, the target downlink SE of 3GPP increases 10 times as the system evolves from GSM to LTE [2]. On the other hand, there is an increasing interest on the environment impact of cellular networks which led to the concept of energy-efficiency defined as the ratio of the spectral-efficiency to the consumed energy. EE is the metric of interest for wireless networks when energy consumption is taken into account. In general, EE means using less energy to accomplish the same task or using the same energy to accomplish more tasks.

EE was previously avoided by most research works and was not regarded by 3GPP as a significant performance evaluator until very lately. In [2], which emphasizes EE besides SE, this metric has been proposed as a framework for energy-efficient solution for future wireless network design. Unfortunately, EE and SE do not always coincide and even conflict sometimes [5,6]. Hence, studying how to balance EE and SE is important to understand these trade-off points for wireless cellular networks.

Given the expensive and scarce spectrum, achieving high data rate coverage requires efficient resource reuse. Orthogonal frequency-division multiplexing (OFDM) is the accepted technology for delivering this performance in LTE [7]. OFDM has the ability to combat frequency-selective fading. In addition, it provides significant flexibility in scheduling that is quite useful for efficient systems: OFDM enables the allocation of distinct sub-channels in frequency domain (sets of adjacent multiple sub-carriers) to different users. When this allocation is extended to the time domain and the transmission is organized into frames/sub-frames with several OFDM symbols, allowing multiple users to access the spectrum simultaneously, the technique is called orthogonal frequency-division multiple access (OFDMA). OFDMA has been extensively investigated from the SE perspective and proposed for next generation wireless communication systems, such as WiMAX (World-wide Inter-operability for Microwave Access) and the 3GPP LTE [8]. While OFDMA can provide high throughput and SE, its EE has previously not been deeply analyzed. To comply with the green radio communication paradigms, it is necessary for OFDMA to ensure a certain level of EE. This is a recent concern in OFDMA network design.

This study presents an overview of SE-EE trade-offs from the OFDMA wireless system point-of-view. Theoretical bounds on SE-EE trade-off are discussed in the next section. The single cell and multi cell study follows the state-of-the-art of the SE-EE trade-off section. Finally, conclusions are drawn in the last section of this chapter.

2.2 State of the Art

Given an available bandwidth, the act of balancing the achievable rate and associated energy consumption of the system is defined as the SE-EE trade-off. Firstly, we discuss the SE-EE relation in the case of a point-to-point connection. Later we provide an overview of SE-EE in OFDMA system for a single cell.

2.2.1 Point-to-Point AWGN Channel

For a point-to-point transmission, the relation between SE and EE is clearly discussed in [2], assuming additive white Gaussian noise (AWGN) channels. Using their work as a reference, we describe in more detail the relationship curve of such SE-EE trade-offs. To characterize the SE-EE trade-off for point-to-point transmission in additive white Gaussian noise (AWGN) channels, Shannon's capacity formula obviously plays a key role. From Shannon's formula [9], the achievable transmission rate, R, under a given transmit power, P, and system bandwidth, B, is

$$R = B \log_2 \left(1 + \frac{P}{B N_0}\right) \tag{2.1}$$

where N_0 is the noise spectral density of AWGN (the value is -174 dBm/Hz).

The maximum achievable transmission rate is the channel capacity of a system which is measured in spectral efficiency. The SE has originally been defined as the ratio of the transmission rate (bits per second) to the spectrum

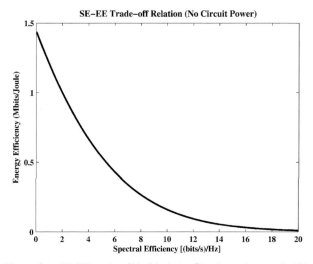

Figure 2.1 SE-EE trade-off in ideal AWGN channel scenario [2].

bandwidth (Hertz) [10, 11]. According to the definition of SE, that is, how efficiently a limited frequency spectrum is utilized and is usually expressed in terms of (bits/s)/Hz. So, mathematically we can express SE as

$$SE = \log_2 \left(1 + \frac{P}{BN_0}\right) \tag{2.2}$$

EE has first been introduced in [12] and is simply defined as the ratio of the capacity to the rate of signal power, i.e. the number of bits that can be transmitted per unit energy of energy consumed. For example, in a channel with power P watts and achievable transmission rate R, the EE can be expressed as

$$EE = \frac{R}{P} = B \log_2 \left(1 + \frac{P}{BN_0}\right) / P \tag{2.3}$$

EE is expressed as bits-per-Joule capacity (bits/Joule).[2] For a high capacity system, when a large number of bits are transmitted, for the sake of convenience EE can be expressed as kilobits-per-Joule (Kbits/Joule), megabits-per-Joule (Mbits/Joule) and so on. Using equations (2.2) and (2.3)

---

[2] $EE = \frac{\text{bits/Second}}{\text{Watt}} = \frac{\text{bits}}{\text{Second} \times \text{Watt}} = \frac{\text{bits}}{\text{Joule}}$.

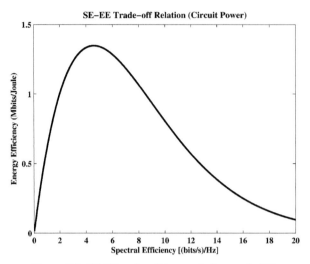

Figure 2.2 SE-EE trade-off in practical scenario [2].

we can derive the SE-EE relation as follows

$$EE = \frac{B \log_2 \left(1 + \frac{P}{BN_0}\right)}{P} = \frac{BSE}{P} = \frac{BSE}{(2^{SE} - 1)BN_0} = \frac{SE}{(2^{SE} - 1)N_0} \tag{2.4}$$

The relation is depicted in Figure 2.1.

The SE-EE relation is monotonically decreasing [2]. The EE tends to converge to a constant when the SE tends to zero [2]; that is

$$\lim_{SE \to 0} EE = \lim_{SE \to 0} \frac{SE}{(2^{SE} - 1)N_0} = \frac{1}{N_0 \ln 2} \tag{2.5}$$

On the other hand, the EE tends to zero when the SE tends to infinity; that is

$$\lim_{SE \to \infty} EE = \lim_{SE \to \infty} \frac{SE}{(2^{SE} - 1)N_0} = 0 \tag{2.6}$$

Therefore, we have the closed-form expression of the SE-EE trade-off for point-to-point transmission under ideal AWGN channel case scenario. However, in reality, the SE-EE trade-off relation is not as simple as the above-mentioned scenario. In particular, under practical concerns apart from the transmission power, circuit wasted power is also a component in the energy consumption. Therefore, and including circuit power as well, EE is defined as transmitted bits per unit energy consumption at the transmitter side, where the energy consumption includes transmission energy consumption and circuit energy consumption. More precisely, if circuit power is considered, the SE-EE curve is a bell shape curve which cannot be expressed in closed-form [13] as depicted in Figure 2.2.

2.2.2 Wireless OFDMA System

The above-mentioned SE-EE trade-off relation is only adequate for modeling a point-to-point transmission, and not a multi user network. Orthogonal frequency-division multiplexing (OFDM) is the accepted technology for delivering this performance in future generation networks. OFDMA offers a quite flexible framework for radio resource management (RRM) as it allows allocation of different portions of radio resources to different users in both the frequency and time domains. In this section we overview the SE-EE trade-off for OFDMA networks, and delve into more details later in this chapter.

In next-generation wireless communication systems, such as WiMAX and the 3GPP-LTE, OFDMA has been widely investigated from the SE viewpoint. In OFDMA, system resource, such as subcarriers and transmit power, needs to be properly allocated to different users to achieve high performance. Allocation of system resources to trade-off SE and EE efficiently for OFDMA network is not in itself a simple task. Figure 2.3 illustrates the resource allocation of a downlink OFDMA network, where subcarriers and power are allocated based on users' channel state information (CSI) and quality of service (QoS) requirements by the base station (BS).

The SE-EE relation for downlink OFDMA networks is quite complex. According to Bohge et al. [14], rate adaptation (RA), which maximizes throughput, and margin adaptation (MA), which minimizes total transmit power, are the two main resource allocation schemes which are commonly used. Therefore, RA aims at SE while MA targets transmit power efficiency. However, neither of them is necessarily energy-efficient. While OFDMA can provide high throughput and spectral-efficiency, its energy consumption is sometimes large. EE and SE do not inevitably agree and a trade-off is required. EE and throughput efficiency can be balanced, e.g. in the uplink, according to user QoS demands and availability of battery power [6].

While SE can always be improved by increasing transmit power in an interference free environment, Miao et al. [6] shows that this does not hold in interference limited communication scenarios since increased transmit power also brings higher interference to the network. On the other hand, conservat-

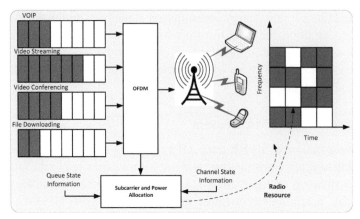

Figure 2.3 Energy-aware resource management in OFDMA [6].

ive energy-efficient communications reduce interference to other users and thus improve overall network spectral-efficiency.

In [15] an energy-efficient design in multi cell scenarios with inter-cell interference for uplink is studied. As shown there, energy-efficient power distribution not only boosts system EE but also refines the SE-EE trade-off due to the conservative nature of power allocation, which sufficiently restricts interference from other cells and improves network throughput.

According to Li et al. [16] there is at least a 15% reduction in energy consumption when frequency diversity is exploited. For uplink transmission with flat fading channels [15], e.g., it is demonstrated that, by applying adaptive modulation the EE increases as the user moves in the direction of the BS, and the nearer the user is to the BS, the higher the modulation order is in the link adaptation.

2.3 Single Cell Spectral Efficiency-Energy Efficiency

A single cell network is the first step to understand a system, whether it is OFDMA or CDMA or TDMA. From a single cell evaluation we get an idea about the trend of a link level which is used as a benchmark for a whole system. This section organization and content is inspired by [13]. In this section, the fundamental trade-off between SE and EE in downlink OFDMA networks is addressed. To get a clear idea about the SE-EE trade-off relation a general framework is needed. Cong et al. [13] provide a generic framework for that trade-off relation. We discuss the trend of the SE-EE curve in single

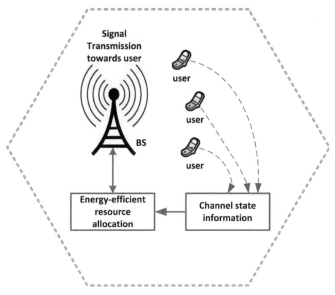

Figure 2.4 Single cell OFDMA network [17].

cell scenario and discuss the impact of channel power gain and circuit power on the SE-EE relation. Finally, we discuss the bounds of the SE-EE curve, which is hard to achieve in a closed form expression [13].

2.3.1 System Model Description

We assume that the base station (BS) is placed in the center on a regular hexagonal grid, as shown in Figure 2.4. Here we consider a standard downlink OFDMA system where a single BS is transmitting data towards a number of users utilizing a number of orthogonal subcarriers. The number of users and the subchannels are denoted by $\mathcal{U} = 1, \ldots, U (u \in \mathcal{U})$ and $\mathcal{S} = 1, \ldots, S(s \in \mathcal{S})$ respectively. We assume that: one subcarrier is exclusively assigned to at most one user to avoid interference among different users, the subcarrier frequency spacing is wide enough and inter-subcarrier interference (ISI) can be ignored, subcarrier allocation which is aligned with the 3GPP LTE standard and related literature [8, 13].

The BS, as well as all users, is assumed to be equipped with a single antenna. It has been assumed that each subcarrier which belongs to a particular user is under frequency-selective fading, for example using ITU Pedestrian-B model. The channel information, i.e., CSI, is sent to the BS over a feedback channel from each user. Furthermore, the BS has perfect channel knowledge for each user instantly. Based on this information, the BS allocates a set of subcarrier to each user and decides on the number of bit in each subcarrier. It is assumed that each subcarrier is exclusively allocated to one user, i.e., sharing a subcarrier by two or more users is not allowed at any given time.

To elaborate, we consider total bandwidth B is equally divided into S subcarrier, each with a bandwidth of $W = B/S$. Using Shannon's formula, we define maximum achievable data rate for a OFDMA system of user u on subcarrier s is [13]

$$r_{u,s} = W \log_2 \left(1 + \frac{p_{u,s} g_{u,s}}{N_0 W} \right) \tag{2.7}$$

where $p_{u,s}$, $g_{u,s}$ and N_0 defines the transmit power, the channel power gain of user u on subcarrier s and the noise spectral density respectively. Therefore, the total system throughput (R) and the total transmit power(P_T) for a single cell downlink OFDMA system is

$$R = \sum_{u=1}^{U} \sum_{s=1}^{S} r_{u,s} \text{ and } P_T = P = \sum_{u=1}^{U} \sum_{s=1}^{S} p_{u,s} \tag{2.8}$$

Eventually, we define the EE as transmitted bits per unit energy consumption at the transmitter side, where the total energy consumption includes total transmit power and circuit power, measured in bits/Joule, and SE as the total transmitted bits per unit spectrum, measured in bits/Hertz. For a downlink OFDMA network, EE and SE are

$$EE = \frac{R}{P_{Total}} = \frac{R}{P_T + P_c} \quad \text{and} \quad SE = \frac{R}{B} \tag{2.9}$$

where P_T and P_c is the transmit power and circuit power respectively. The EE is thus expressed as

$$EE = \frac{\sum_{u=1}^{U} \sum_{s=1}^{S} r_{u,s}}{\sum_{u=1}^{U} \sum_{s=1}^{S} p_{u,s} + P_c} \tag{2.10}$$

2.3.2 SE-EE Trade-off Relation

To understand the SE-EE trade-off it is important to understand the shape of the SE-EE curve. If we plot equation (2.10) for any given SE the shape of

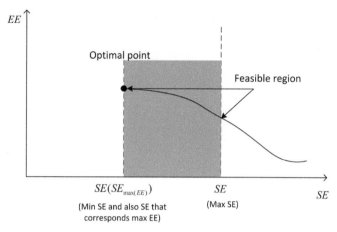

Figure 2.5 SE-EE in downlink OFDMA, scenario 1 [13].

the curve would be like quasiconcave, which means EE is quasiconcave in SE. Cong et al. [13] provide a list of theorems to illustrate this idea. Xiong et al. [13] provides a theorem exists on the relation between SE-EE. For any given SE, EE is strictly quasiconcave in SE if there are a sufficiently large number of subcarriers and a solution for the traffic allocation. Moreover, in the SE region (from minimum SE to maximum SE), the EE provides three types of scenarios in quasiconcavity, as explained hereafter. The theorem demonstrates the quasiconcavity of EE on SE and reconfirms the existence and the uniqueness of the globally optimal EE as illustrated in Figures 2.5, 2.6 and 2.7. This approach gives us a uniform SE (or total rate) perspective rather than that of a vector of split user rates, which makes it easier to track the SE-EE relation.

Three possible scenarios for the SE-EE curve are discussed in [13] as depicted in Figures 2.5, 2.6 and 2.7. In scenario 1, the optimal point of SE-EE trade-off, i.e. the point of the optimal EE is reached when the point of the optimal SE equals the point of minimum SE, as illustrated in Figure 2.5. In scenario 2, the optimal EE point is reached when the point of the optimal SE equals the point of maximum SE i.e. the trade-off point (optimal point of EE) is found on the point where the point of maximum SE is reached as illustrated in Figure 2.6. In scenario 3, the point of optimal EE is reached when optimal SE equals $SE_{\max(EE)}$ ($SE_{\max(EE)} \rightarrow$ the SE that corresponds the maximum EE) as sketched in Figure 2.7. We should keep in mind that the point of the optimal SE is the optimal SE of equation (2.9).

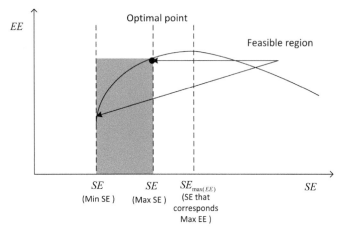

Figure 2.6 SE-EE in downlink OFDMA, scenario 2 [13].

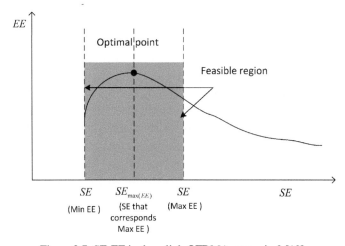

Figure 2.7 SE-EE in downlink OFDMA, scenario 3 [13].

Note that the SE-EE trade-off points such as the optimal EE point equals maximum EE point in scenarios 1 and 3 while it is smaller than the latter in scenario 2. In addition, this SE-EE trend offers a simple and adaptive way to determine a desirable and feasible data transmission rate, in practical situations [13]. For simplicity, we should assume data transmission is appropriately predetermined.

To have more insight into the SE-EE trade-off relation and to understand the impact of the channel power gain and circuit power on the SE-EE relation some of the properties based on the above mentioned work are given verbatim in the following [13]:

Property 1 *For any fixed SE, the EE is non-decreasing with channel power gain. Consequently, both the optimal EE and the global maximum EE are non-decreasing with channel power gain* [13].

Property 2 *For any fixed SE, the EE is strictly decreases with circuit power. Consequently both the optimal EE and the global maximum EE strictly decrease with the circuit power, P_c. Besides, the optimal SE is non-decreasing with the static power* [13].

Property 3 *The optimal EE is not necessarily achieved at minimum SE even if circuit power is zero* [13].

2.3.3 Bounds on the SE-EE Curve

In spite of the known trends of the SE-EE curve, it is very difficult to find the exact solution of the upper bounds and lower bounds of the curve. As it is extensively studied in [13], Lagrange Dual Decomposition (LDD) [18] is used to find the bounds of the SE-EE curve, an approach that has been used in [19, 20] for similar OFDMA allocation problems. Since our goal is to give an overview to the reader and this allocation process is out of the scope, we avoid in-depth mathematical analysis here.

The water-filling algorithm [21] is used to allocate subcarriers for each user so that we have an idea about the bounds of the SE-EE trade-off. Using the water-filling algorithm the achievable upper bound of SE-EE curve on the minimum transmit power is found. This method is completely described in [13]. On the other hand, a lower bound on the minimum transmit power can be obtained by relaxing the channel allocation.

Interestingly, although it is hardly possible to derive a closed-form expression for the upper and lower bound of the curve, the differentiation of equation (2.10) with respect to SE can give us some idea, because the differentiation of that function can easily and accurately determine its sign to solve bounds on the transmit power minimization. This coincides with and relates to theorem 4 of [13].

In this section, we have presented the SE-EE relation in a single cell downlink OFDMA network, which is important for designing Green Wireless Communications that require a better balance between SE and EE. The SE-EE relation is shown to be a quasiconcave function with the help of the ground breaking research article [13], and the impact of channel power gain and circuit power is also discussed. Methods for finding a lower and upper bound on the SE-EE curve are summarized in a comprehensive way avoiding complex mathematical analysis. Similar theoretical approaches will be used in the next section for multi cell OFDMA networks.

2.4 Multi Cell Spectral Efficiency-Energy Efficiency

Miao et al. [22] have extensively studied energy-efficient design in multi cell scenarios with inter-cell interference for uplink OFDMA system. They reported that due to the conservative nature of the power allocation, SE-EE trade-off is reduced using energy-efficient power distribution which also increases the system EE. That means the SE-EE trade-off relation is able to restrict interference from other cells and improves network throughput. Since the above mentioned research activity is based on uplink there is still a lot to ponder about downlink multi cell OFDMA system. So far very few works have been done in downlink.

For the multi cell OFDMA system the EE is a function which includes interference from other cells and co-channel interference is considered. The trade-off between SE and EE in this system mainly depends on accurate channel state information at the transmitter (CSIT) and feedback power consumption. Analyzing the EE under practical the CSIT model provides more practical significance, and can lead to insights in the SE-EE trade-off relation. Existing work on the above topic is relatively limited, since even the SE of the multi cell multi user systems is still under investigation.

2.4.1 System Model

In this section, we describe the system model of a generic multi cell OFDMA downlink system. For example, we consider a multi cell OFDMA system: the overall frequency bandwidth B is divided into S subcarriers/subchannels each with a bandwidth of $W = B/S$ and all the subchannels are reused among C distinct cells. In each cell, the unique BS is located at the center of the cell, and the served user terminals are randomly spread throughout the service area. From now on we use the terms BS and cell interchangeably.

The propagation channel is frequency selective and slowly time-varying, such that each BS is able to exchange CSI with the served user terminals via a dedicated feedback channel. If we consider a system consisting of one tier of six cells with a central cell as a starting point, then, as a result, all the

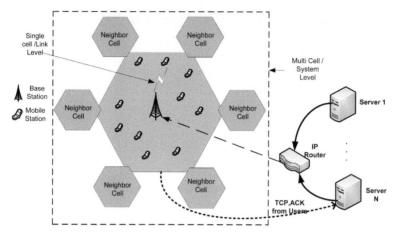

Figure 2.8 Multi cell OFDMA network.

transmitted signals are collected from the central cell, with the other six cells serving as interferers. A system with 19 cells has then six cells in tier two, whereas the other 12 cells constitute tier three, with the central cell treated as tier one. With this approach the increase in the cell number leads to more tiers in the system. Figure 2.8 illustrates a multi cell scenario of a downlink OFDMA system.

The major difference between a single cell and multi cell system is the interference which occurs in a multi cellular system. Generally, interference acts as a flat fading signal from the interfering base stations. Interference from BS to BS also termed as a co-channel interference. Thus we shall apply equation (2.7) with signal-to-noise-ratio (SNR $= \frac{p_{u,s}g_{u,s}}{N_0 W}$) replaced by the signal-to-interference-plus-noise-power-ratio (SINR $= \frac{p_{u,s}g_{u,s}}{I+N_0 W}$).

For example, we define $C = \{1, \ldots, C\}$ (for any c, $c \in C$) as the set of cells/BSs in the system, and also the set of the number of users and the subcarriers are denoted by $U = \{1, \ldots, U\}$ (for any u, $u \in U$) and $S = \{1, \ldots, S\}$ (for any s, $s \in S$) respectively. For the downlink cases, due to the co-channel interference from other cells, the output SINR at the receiver for the user u user on subcarrier s from the BS c is defined in the following equation:

$$\text{SINR} = \frac{p_{u,s}^c g_{u,s}^c}{I + N_0 W} \tag{2.11}$$

where $p_{u,s}^c$, $g_{u,s}^c$ and N_0 defines the transmit power and the channel power gain of user u on subcarrier s from the BS c and the noise spectral density respectively. I denotes the interfering signal power from the neighboring BSs. We can elaborate the value of I as $I = \sum_{i=1, i \neq c}^{C} p_{u,s}^{i}$. After substituting the value of I in equation (2.11), we get

$$\text{SINR} = \frac{p_{u,s}^{c} g_{u,s}^{c}}{\sum\limits_{i=1, i \neq c}^{C} p_{u,s}^{i} + N_0 W} \tag{2.12}$$

2.4.2 Bounds on the SE-EE Curve

Using Shannon's formula [9], the maximum achievable data rate for a multi cell OFDMA system of user u on subcarrier s from the BS c is

$$r_{u,s} = W \log_2 \left(1 + \frac{p_{u,s}^{c} g_{u,s}^{c}}{I + N_0 W} \right)$$

$$r_{u,s} = W \log_2 \left(1 + \frac{p_{u,s}^{c} g_{u,s}^{c}}{\sum_{i=1, i \neq c}^{C} p_{u,s}^{i} + N_0 W} \right) \tag{2.13}$$

According to the definition of EE from the previous section, we define the EE for a multi cell scenario as

$$EE = \frac{\sum_{u=1}^{U} \sum_{s=1}^{S} r_{u,s}^{c}}{\sum_{u=1}^{U} \sum_{s=1}^{S} p_{u,s}^{c} + P_c} \quad \text{(for any } c; c \in \mathcal{C}) \tag{2.14}$$

Due to interference-limited scenario in multi cell system, the EE decreases in the multi cell compared to the single-cell case as illustrated in Figure 2.9. From this figure we can get the same quasiconcave trend of SE-EE relation i.e., EE is quasiconcave in SE.

Very little research has been done for SE-EE trade-off in multi-cell OFDMA system. Karray [23] provides some insight of the SE-EE bound although it does not consider complete practical channel propagation model such as frequency-selective fading channel. Karray [23] concludes with a proposition which demonstrates that the EE tends to zero when the transmitted power tends either to zero or infinity. Eventually, they clarify EE admits a maximum for some non-null and a finite value of the transmitted power in terms of SE; the value of the optimal power depends on the configuration of the network. More interested readers can refer to [23] to understand the proof. However, the SE-EE relation for general multi cell networks, including downlink OFDMA network, is a subject that still needs extensive research.

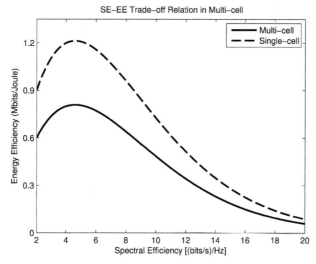

Figure 2.9 SE-EE trade-off relation between multi cell and single cell.

2.5 Significance of Circuit Power in SE-EE Trade-Off

In this section we focus on some of the results we achieved through our investigation. As we already mentioned in the previous section, EE is defined as transmitted bits per unit energy consumption at the transmitter side, where the energy consumption includes transmission energy consumption and circuit energy consumption of transmitter in the active mode.

$$EE = \frac{R}{P} = \frac{R}{P_{\text{Total}}} = \frac{R}{P_{\text{Transmit}} + P_{\text{Circuit}}} \qquad (2.15)$$

In the transmit/active mode of transmitter, and besides transmit power, the energy consumption also includes circuit energy consumption that is incurred by signal processing and active circuit blocks, such as analog-to-digital converter, digital-to-analog converter, synthesizer, and mixer [23].

From [24, 25], circuit power can be divided into two parts. One is static circuit power which is also a fixed entity; whilst the other one is dynamic circuit power consumption which is a dynamic circuit power factor in terms of per unit data rate. Hence the circuit power is modeled as follows [25]:

$$P_{\text{Circuit}} = P_{\text{Static}} + P_{\text{Dynamic}} = P_{\text{Static}} + \rho R \qquad (2.16)$$

where ρ [25] is a constant denoting dynamic power consumption per unit data rate (R). Figure 2.10 demonstrates SE-EE curve of an LTE-advanced down-

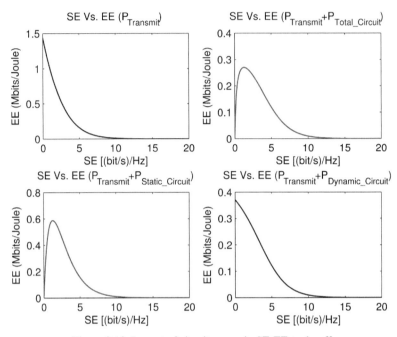

Figure 2.10 Impact of circuit power in SE-EE trade-off.

link system. From this figure we can attain SE-EE trade-off relation while using different power consumption models. As we already know, in a ideal scenario EE always decreases in terms of SE, while in practical scenario EE is quasiconcave in SE. The demonstration of this result is given in Figure 2.10.

The SE-EE curve in transmit power and transmit-plus-dynamic-circuit power shows the similar trend. The reasoning behind this trend is that the dynamic circuit power factor actually scales the data transmission rate which has no relation to any static circuit power factor. Therefore, the trend remains the same as in the ideal SE-EE scenario. On the other hand, the SE-EE curve trend is quasiconcave in terms of both the total circuitry power consumption and the static circuitry power consumption. From Figure 2.10, it is evident that the static circuit power plays a pivotal role the quasiconcave nature of the curve. The reason being is that the static circuit power has no coupling with the transmission rate (as well as spectral-efficiency or bandwidth). This type of circuit power is responsible for the quasi-concave trend of the curve; not the dynamic circuit power. So if we use only static circuit power without considering dynamic circuitry consumption, the SE-EE curve trend remains

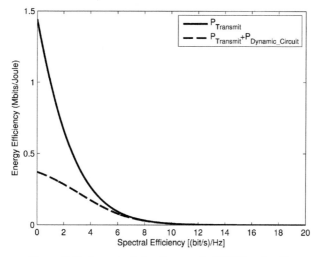

Figure 2.11 Impact of circuit power in SE-EE trade-off.

quasiconcave like the SE-EE curve trend using the total circuit power (both static and dynamic circuitry consumption).

In Figures 2.11 and 2.12 we compare EE amongst the different types of energy consumption such as using only transmission power, transmission power-plus-only dynamic circuit power, transmission power-plus-total circuit power and transmission power-plus-only static circuit power.

The energy-efficiency decreases when consumed power includes transmit power and dynamic circuit power factor as illustrated in Figure 2.11. That means dynamic circuit power reduces the EE. In the case of comparing EE between total power (which includes both transmit power and total circuit power) and the power considering transmit-plus-static circuit power we find that EE decreases in the latter much more aggressively, as demonstrated in Figure 2.12. It is fairly evident that to have a SE-EE balance, circuit power plays a major role towards shaping this trade-off.

2.6 Conclusions

In this chapter, we presented an overview of the SE-EE trade-off in OFDMA wireless cellular networks. Current research trends are focused on green communications which consists in maximizing the number of bits per unit energy consumption that can be reliably conveyed through the channel. In a point-

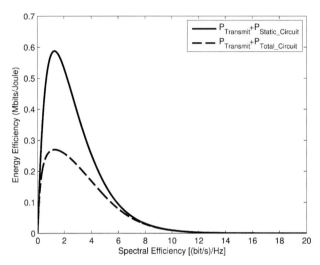

Figure 2.12 Impact of circuit power in SE-EE trade-off.

to-point link, the SE-EE relation is straightforward as far as the total power is concerned. When OFDMA is assumed we have a more challenging scenario with greater coupling on the network's energy-efficiency. However, these conclusions do not hold necessarily in practical scenarios where the circuitry energy is also considered [13].

In this chapter, we have discussed a framework for SE-EE trade-off with the support of the existing literature [23], which should be at the forefront of system design for more energy compliant networks. We have shown that, in practical systems, the trade-off relations usually deviate from the simple monotonic curves derived from Shannon's formula as summarized in Section 2.2. Moreover, most of the existing literature mainly focuses on the point-to-point single cell case. Therefore, the SE-EE trade-off relations under more realistic and complex network scenarios deserve further investigation. The insights, such as how to improve the SE-EE trade-off curves from a holistic view and how to tune the operating point on the curve to balance the specific system requirements, are expected to guide practical system designs toward more energy conscious networks, and define the next step in Green Communication research.

2.7 Summary

Since SE and EE are two important system performance indicators, the trade-off between EE and SE for general OFDMA networks should be exploited to guide system design according to the existing literature. The bounds and achievable SE-EE regions for downlink OFDMA networks are important for the system designer [16]. Meanwhile, proper utility functions should be investigated for locating the optimum operating point on the boundary of SE-EE region [2]. The current state-of-the-art of the existing research demonstrated that the larger EE can be achieved through energy-efficient design [16] and that will be applied for next generation techniques such as MIMO [26], coordinated base station system [27] and relay transmission [28]. However, most work is still in the initial stage, and more effort is needed to carry out the investigation of the SE-EE trade-off.

A better balance between EE and SE is required. With the help of the existing literature [23], the SE-EE trade-off is defined to be a quasiconcave function and the impact of the channel power gain and circuit power is analyzed. Some of the issues such as imperfect channel state information with average per-user rate requirement have not yet been tackled.

Optimization of resources such as power and bandwidth are needed to ensure energy-efficient OFDMA wireless system. Advances in the layering framework in wireless protocol, for example, interdependencies of the different layers such as cross-layer resource optimization, are required to determine the fundamental bounds on achievable energy-efficiency. Correspondingly, research into practical realization and hardware implementations of energy-efficient protocols should be explored.

In this chapter, an accurate closed-form approximation of the SE-EE trade-off has not been discussed for multi cell scenario as very little research has been carried out so far; hence existing research of the SE-EE trade-off can be used as a starting point. Lastly, the next evolutionary stage of mobile networks is LTE-Advanced, that consider several deployment scenarios also based on OFDMA access such as CoMP (Coordinated Multi Point), Femto Cell, and SON (Self Organizing Network) that should be researched further to find the SE-EE trade-off for ensuring energy saving and hence reducing the operator's operational expenditure.

Acknowledgements

The authors would like to acknowledge the project No. 23205 – GREEN-T, co-financed by the European Funds for Regional Development (FEDER) by COMPETE – Programa Operacional do Centro (PO Centro) of QREN.

Kazi Huq, PhD student at the University of Aveiro, would like to acknowledge the support of the grant of the Fundacão para a Ciência e a Tecnologia (FCT, Portugal), with reference number SFRH/BD/62165/2009.

References

[1] M. De Sanctis, E. Cianca, and V. Joshi, Energy efficient wireless networks towards green communications, *Wireless Personal Communications*, vol. 59, no. 3, pp. 537–552, February 2011. [Online] available from http://www.springerlink.com.miman.bib.bth.se/content/e764hq2h2h83081t/.

[2] Y. Chen, S. Zhang, S. Xu, and G. Y. Li, Fundamental trade-offs on green wireless networks, *IEEE Communications Magazine*, vol. 49, no. 6, pp. 30–37, June 2011.

[3] 3GPP, Technical specification group radio access network; evolved universal terrestrial radio access (E-UTRA); Potential solutions for energy saving for E-UTRAN (Release 10), 3GPP TR 36.927,V10.0.0, Technical Report, September 2011.

[4] T. Chen, Y. Yang, H. Zhang, H. Kim, and K. Horneman, Network energy saving technologies for green wireless access networks, *Wireless Communications, IEEE*, vol. 18, no. 5, pp. 30–38, October 2011.

[5] F. Meshkati, H. V. Poor, and S. C. Schwartz, Energy-efficient resource allocation in wireless networks, *IEEE Signal Processing Magazine*, vol. 24, no. 3, pp. 58–68, May 2007.

[6] G. Miao, N. Himayat, Y. Li, and A. Swami, Cross-layer optimization for energy-efficient wireless communications: A survey, *Wirel. Commun. Mob. Comput.*, vol. 9, no. 4, pp. 529–542, 2009.

[7] S. Parkvall and D. Astely, The evolution of LTE towards IMT-Advanced, *Journal of Communications*, vol. 4, no. 3, April 2009. [Online] available from http://ojs.academypublisher.com/index.php/jcm/article/view/0403146154.

[8] I. F. Akyildiz, D. M. Gutierrez-Estevez, and E. C. Reyes, The evolution to 4G cellular systems: LTE-Advanced, *Physical Communication*, vol. 3, no. 4, pp. 217–244, 2010.

[9] Claude Elwood Shannon, A mathematical theory of commuication, *The Bell System Technical Journal*, vol. 27, pp. 379–423, 623–656, 1948.

[10] D. N. Hatfield, Measures of spectral efficiency in land mobile radio, in *Proceedings of 25th IEEE Vehicular Technology Conference, 1975*, vol. 25. IEEE, pp. 23–26, January 1975.

[11] W. Lee, Spectrum efficiency in cellular [radio], *IEEE Transactions on Vehicular Technology*, vol. 38, no. 2, p. 69–75, 1989.

[12] Hyuck M. Kwon and Theodre G. Birdsall, Channel capacity in bits per Joule, *IEEE Journal of Oceanic Engineering*, vol. 11, no. 1, pp. 97–99, January 1986.

[13] C. Xiong, G. Y. Li, S. Zhang, Y. Chen, and S. Xu, Energy- and spectral-efficiency tradeoff in downlink OFDMA networks, *IEEE Transactions on Wireless Communications*, vol. 10, no. 11, pp. 3874–3886, November 2011.

[14] M. Bohge, J. Gross, A. Wolisz, and M. Meyer, Dynamic resource allocation in OFDM systems: An overview of cross-layer optimization principles and techniques, *IEEE Network*, vol. 21, no. 1, pp. 53–59, February 2007.

[15] G. Miao, N. Himayat, Y. Li, and D. Bormann, Energy efficient design in wireless OFDMA, in *Proceedings of IEEE International Conference on Communications 2008 (ICC'08)*, pp. 3307–3312, May 2008.

[16] G. Li, Z. Xu, C. Xiong, C. Yang, S. Zhang, Y. Chen, and S. Xu, Energy-efficient wireless communications: Tutorial, survey, and open issues, *IEEE Wireless Communications*, vol. 18, no. 6, pp. 28–35, December 2011.

[17] A. Akbari, R. Hoshyar, and R. Tafazolli, Energy-efficient resource allocation in wireless OFDMA systems, in *Proceedings of 2010 IEEE 21st International Symposium on Personal Indoor and Mobile Radio Communications (PIMRC)*. IEEE, pp. 1731–1735, September 2010.

[18] M. Chiang, S. Low, A. Calderbank, and J. Doyle, Layering as optimization decomposition: A mathematical theory of network architectures, *Proceedings of the IEEE*, vol. 95, no. 1, pp. 255–312, January 2007.

[19] K. Seong, M. Mohseni, and J. M. Cioffi, Optimal resource allocation for OFDMA downlink systems, in *2006 IEEE International Symposium on Information Theory*. IEEE, pp. 1394–1398, July 2006.

[20] X. Zhou, G. Y. Li, D. Li, D. Wang, and A. C. Soong, Probabilistic resource allocation for opportunistic spectrum access, *IEEE Transactions on Wireless Communications*, vol. 9, no. 9, pp. 2870–2879, September 2010.

[21] R. Cheng and S. Verdu, Gaussian multiaccess channels with ISI: Capacity region and multiuser water-filling, *IEEE Transactions on Information Theory*, vol. 39, no. 3, pp. 773–785, 1993.

[22] G. Miao, N. Himayat, G. Li, A. Koc, and S. Talwar, Interference-aware energy-efficient power optimization, in *Proceedings of IEEE International Conference on Communications 2009 (ICC'09)*, pp. 1–5, June 2009.

[23] M. K. Karray, Spectral and energy efficiencies of OFDMA wireless cellular networks, in *Proceedings of Wireless Days (WD), 2010 IFIP*. IEEE, pp. 1–5, October 2010.

[24] O. Arnold, F. Richter, G. Fettweis, and O. Blume, Power consumption modeling of different base station types in heterogeneous cellular networks, in *Proceedings of Future Network and Mobile Summit, 2010*. IEEE, pp. 1–8, June 2010.

[25] C. Isheden and G. P. Fettweis, Energy-efficient multi-carrier link adaptation with sum rate-dependent circuit power, in *Proceedings of 2010 IEEE Global Telecommunications Conference (GLOBECOM 2010)*. IEEE, pp. 1–6, December 2010.

[26] S. Cui, A. Goldsmith, and A. Bahai, Energy-efficiency of MIMO and cooperative MIMO techniques in sensor networks, *IEEE Journal on Selected Areas in Communications*, vol. 22, no. 6, pp. 1089–1098, August 2004.

[27] O. Onireti, F. Heliot, and M. Imran, On the energy efficiency-spectral efficiency trade-off in the uplink of comp system, *IEEE Transactions on Wireless Communications*, vol. 11, no. 2, pp. 556 –561, February 2012.

[28] R. Madan, N. Mehta, A. Molisch, and J. Zhang, Energy-efficient cooperative relaying over fading channels with simple relay selection, *IEEE Transactions on Wireless Communications*, vol. 7, no. 8, pp. 3013–3025, August 2008.

3

Context Based Node Discovery Mechanism for Energy Efficiency in Wireless Networks

Muhammad Alam[1,2], Michele Albano[3], Ayman Radwan[1] and
Jonathan Rodriguez[1]

[1]*Instituto de Telecomunições, Aveiro, Portugal*
[2]*Universidade de Aveiro, Aveiro, Portugal*
[3]*CISTER/ISEP Polytechnic Institute of Porto, Portugal*
e-mail: alam@av.it.pt

Abstract

Energy has now become a vital resource in wireless networks. With the introduction of high data rate wireless technologies, advanced wireless devices will be able to support an ever growing demanding for high resource consumption applications (in terms of processing power as well as energy requirement for data transmission), e.g. interactive games, video sharing, etc., and as a result the energy saving mechanisms will be at the forefront of system design. The mobile devices with hybrid character, containing ad hoc and infrastructure access, will need mechanisms that keep them connected in an energy efficient way. Wireless Personal Area Networks (WPANs) have gained strong interest for energy compliant communications, especially their application with cooperative and context-aware protocols. These two paradigms working in synergy have produced significant energy reduction to the node/network discovery mechanism which is seen as one of the highest energy consumers in your mobile device. In this chapter, we detail the latest innovations on lightweight node discovery mechanisms for WPANSs. In the first instance we identify the benchmark performance, as it is today; in that the node to be connected to either a cluster or a solitary node, knowledge of the neighboring devices are required to choose a suitable node for cooper-

*Shahid Mumtaz and Jonathan Rodriguez (Eds.), Green Communication in 4G
Wireless Systems,* 33–52.

ative communication which requires exhaustive search resulting in the radio interface being always "on". Moreover, we advance this by proposing new energy efficient discovery mechanisms that are essential for latching onto the benefits of cooperative mechanims. We present a context based beaconing mechanism both for node discovery and cooperative cluster formation. We exploit the available beacon Information Elements (IEs) of UWB and have proposed a new Context Extraction Function (CxEF). Beyond this, we also present a detailed representation of bit mapping for energy and cooperation in beacons. The chapter is concluded by simulation results, which demonstrate the potential energy saving that can be achieved using the proposed node discovery algorithm based on Context Awareness.

Keywords: Node discovery, Wireless Personal Area Networks, Ultrawide Band, PicoNet, energy efficiency.

3.1 Introduction

Energy has now become a vital resource in wireless networks. With the introduction of high data rate wireless technologies, advanced wireless devices will be able to support an ever growing demanding for high resource consumption applications, in terms of processing power as well as energy requirement for data transmission. Therefore, energy is a critical resource in battery-powered mobile device and dissipates quickly to keep mobile devices connected to the network over extended periods of time [1]. To trim down the energy consumption of mobile devices, there is a need for a cross-layer design that aims to maximize the energy consumption from a holistic perspective. A plethora of international research projects currently addresses green communications, in particular C2POWER [2], and GREEN-T [3] target energy reduction on the "mobile side" by investigating energy efficient networking topologies. These projects investigate how short-range communications working in synergy with long-range exploiting a context-aware overlay network can lead to significant power savings of up to 50%. However, this energy saving target is viable as long we can reduce the node discovery procedure in short-range networks when we are scanning for new piconet members or so called "cooperative" mobile subscribers, the latter term is often referred to in [2, 3]; the node/network discovery mechanism which is seen as one of the highest energy consumers in your mobile device. In this chapter, we innovate on the latest in lightweight node discovery mechanisms for WPANs. However, to appreciate fully the energy saving that can be achieved using context-aware devices; there is a need to understand how node discovery mechanisms work in today's WPAN radios, and how "energy hungry" they really are.

The node discovery process is an initial step to start a cooperative communication network. Most of the node discovery protocols and mechanisms are based on the conventional scanning, which is expensive in terms of energy due to spending long periods of time in the "listening" state, and performing broadcast communication. To conserve energy, MAC protocols with proper sleep/wake duty cycles have been proposed for wireless networks, providing a balance in sleep and wake time. Although suitable duty cycles contribute to conserve energy, keeping a node in sleep mode may weigh down a network or a part of the network [4]. Furthermore, for a solitary node to discover another node or network, even an efficient duty cycle cannot contribute effectively because the node is not aware of the exact spot where to turn on its interface to find a cooperative node. On the other hand, sometimes a node is in the vicinity of a cluster or a node but lacks information about the attributes of the cluster or node to cooperate with.

Keeping these problems in mind, we propose a scenario and a smart scanning mechanism to tackle both the node discovery and the cooperative cluster formation in an energy efficient way. Ultra Wideband (UWB) is a possible candidate as short-range communication technology, since it offers a number of advantages over other short range technologies. Characteristics such as large bandwidth, low power requirements and precise positioning capabilities have made UWB attractive as WPAN technology [5]. Current and future UWB advantages have been presented in [6–8]. UWB MAC channel time is divided into superframes, which are in turn divided into 256 Medium Access Slots (MAS). The two major subdivisions of the superframe are: the beacon period (BP) and the data transfer period (DTP). A beacon frame in each superframe is transmitted by each MT to advertise its presence and its local view of the network. The beacon frame of UWB consists of a number of information elements (IEs) containing the timing and control information [6]. There is a number of available fields in the beacon frames that can be modified or new fields can be proposed to exchange novel kinds of information, e.g. the energy level of an MT. In DTP MTs communicate with each other and the MASs are retained by either prioritized channel access (PCA) or a distributed reservation protocol (DRP), or a combination of both [6]. A detailed study of PCA can be found in [9,10] and of DRP in [11]. Apart from modifying beacon frames, we also created an extraction function that extracts the information, maps the semantic information onto values, and forwards them to specified modules of the node. Furthermore, a detailed representation of the battery level for energy efficient node discovery and cooperative cluster formation is presented.

3.2 Current Node Discovery in Short Range Technologies (WPAN)

Node discovery is a research topic that has been widely investigated, but most current results only apply to Personal Area Networks (PANs). Initially all the nodes are unaware of each other's presence, so node discovery is the initial state for joining and building an ad-hoc network. The process of node discovery is the first step in configuring and optimizing the topology of the network. Given node mobility in PAN and the dynamics of the applications [12], nodes will join or leave the network during application's lifetime, and radio links will be greatly affected. To cope with this dynamicity, the network should be able to quickly reconfigure itself without the user's intervention. Thus, the mechanisms for the discovery of neighbors, the creation of connections, the scheduling of transmissions and the formation and re-configuration of the network topology should be seamless and energy efficient.

Bluetooth (BT) [13] is a short-range wireless communication protocol that can have a coverage distance between 10 and 100 m characterized by a bit rate of up to 2.1 Mbps with low energy consumption attributes. There are several possible ways by which a BT device can interact with other BT devices. In the most basic one, one of the devices acts as master and the others are slaves, and form an ad-hoc network which is called PicoNet. A PicoNet has a star topology, where one master arbitrates the communication between 1–7 active slaves. A PicoNet can have up to 255 slaves, but at most seven of them can be active at one time, while the other devices would be parked (inactive). A parked node can change its status to active when it wants to perform communication and, if the PicoNet already comprises seven slaves, one of the other slaves will be switched to the parked state. The role of the master is also switched to another device when the active master departs the PicoNet. Each of the active slaves has a 3-bit Active Member address (AM_ADDR). All channel accesses are regulated by the master. All devices in a PicoNet share the same frequency hopping sequence by the use of a slotted Time Division Duplex (TDD) technique. The physical channel is sub-divided into time units, known as slots, and data is transmitted between Bluetooth enabled devices in packets that are positioned in these slots [13]. Each other PicoNet

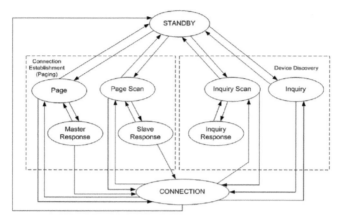

Figure 3.1 State transition diagram for Bluetooth [16].

uses a different Frequency Hopping Succession (FHS) in order to decrease interference with PicoNets located in the vicinity [14, 15]. Multiple PicoNets inter-connect to form a scatternet in order to extend the coverage range of the network. A node that joins two PicoNets is called a bridge node, and while a bridge node can participate in more than one PicoNet, it can typically be active in only one at a time. A node cannot be master in more than one PicoNet, but it can be master in one or zero PicoNets, and slave in more than one PicoNet at a time. Hence, a small number of nodes can take action as bridges and perform the job of relaying packets across multiple PicoNets. In a scatternet, each PicoNet is recognized by its specific FHS. Device discovery and connection establishment are fundamental operations that have to be done to enable communication between two BT enabled devices. In BT the nodes adhere to (1) Start; (2) Node discovery; (3) Synchronization; (4) Paging; and (5) Connection establishment.

The state transition diagram for Bluetooth node discovery is shown in Figure 3.1. The device can switch between a numbers of states. In the STANDBY state the device is not a part of any PicoNet while in the CONNECTION state it is part of a PicoNet. If a device wants to discover a node and to be a part of PicoNet, it can move from STANDBY or CONNECTION state to the INQUIRY state. The device that wants to discover another device to form a PicoNet, is designated as the master of the PicoNet. On the other hand, if a node wants to be discovered by others can move from STANDBY or CONNECTION state to the INQUIRY SCAN state. While a device is in the process of discovery, it can temporarily move from INQUIRY SCAN to the

INQUIRY RESPONSE state, to answer the inquiry packets it has received. The device that is willing to be discovered will be a slave in a PicoNet.

UWB communication systems are usually classified as "any communication system whose instantaneous bandwidth is multiple times greater than the minimum bandwidth required to deliver information". The excess bandwidth is a defining characteristic of UWB. The most prominent protocol of the UWB category is IEEE 802.15.3, whose physical layer adopts Multiband Orthogonal Frequency Division Multiplexing (MB-OFDM) UWB and Direct Sequence-Code Division Multiplex Access (DS-CDMA) UWB technologies. It allows 245 wireless user devices in a range of several centimeters to 100 m to access a network simultaneously, at a top rate of 55 Mb/s, and it provides fixed and mobile equipment with high-rate wireless access at 2.4 GHz. The standard IEEE 802.15.3 specifies five original data rates: 11, 22, 33, 44, and 55 Mb/s, which will influence the transmission distances. For example, the maximum transmission distance at 55 Mb/s is 50 m, while at 22 Mb/s is 100 m. High data rates (such as 55 Mb/s) can support low-delay multi-medium access and big-file transfer service, while low rates such as 11 and 22 Mb/s) offer long-distance connection between audio equipment. IEEE 802.15.3 has all the elements required for QoS guarantee. Moreover, it uses Time Division Multiple Access (TDMA) technology to allocate channels to avoid collisions. IEEE 802.15.3 defines the protocols for the physical and MAC layers. Its MAC protocol is inspired by the IEEE 802.11 MAC protocol for Wireless Local Area Network (WLAN). Therefore, it is based on the Ad hoc structure and still has a trace of the star topology (as illustrated in Figure 3.2). The PicoNet is the basic unit of this kind of UWB network. The core device of the PicoNet is called PicoNet Coordinator (PNC) and is responsible for offering the synchronized clock, QoS control, power saving mode and access control. The PicoNet, which is an ad hoc network, exists only when there is a demand for communication and disappears with the end of the communication. The communication node of a PicoNet is called DEV in UWB terminology. Data transmission is performed between the DEVs, but the control information is exchanged only with PNC.

A PicoNet emerges when the PNC starts to send beacons. Even if there are no communicating nodes, a PNC sending beacons can be regarded as a PicoNet. When a PicoNet is emerging, the PNC first finds a usable channel and sends a beacon frame to guarantee that the channel is free: then the PicoNet is established on this channel. Following this, the PNC role can still be changed by handover control. However, IEEE 802.15.3 does not support the function of integrating two PicoNets into one.

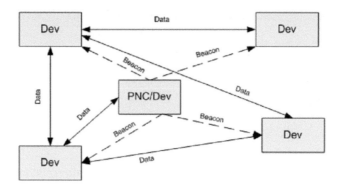

Figure 3.2 IEEE 802.15.3 WPAN architecture [17].

The PNC implements allocation of the wireless resources by sending beacons, which carry network control parameters (such as network synchronization and maximal transmission power), information about allocation of channel time slot, and more. CAP uses the MAC mechanism for CSMA/CA. In the CTAP, allocation may be implemented by basic TDMA. MCTA can also use TDMA to allocate time slots. Additionally, equipments are allowed to share the MCTA (based on the ALOHA protocol [18]).

New protocols for the MAC layer keep emerging. On the one hand, the IEEE 802.15 working group planned to initiate a study on new MAC protocols in IEEE 802.15.3b. On the other hand, MB-OFDM has great support and promotion from equipment vendors and research institutes, and Multiband OFDM Alliance (MBOA) is making its own MAC protocol. In order to better support low-end communication equipment, the MBOA MAC protocol will support both the central control structure and the distributed network topology.

In WiMedia protocol, once a UWB radio has powered up and is ready to begin operation, the first thing to be done is to find a channel to operate. A channel is comprised of a frequency band group which describes the frequency boundaries of operation, and a time frequency code (TFC), which dictates the manner in which hopping is conducted between the frequency bands that make up the band group. The next step is starting to listen to the spectrum in search for someone with whom to talk. Regulatory requirements allow the first device to beacon for a period of up to 10 seconds [19]. The fundamental component of WiMedia MAC communications is the superframe, which is shown in Figure 3.3.

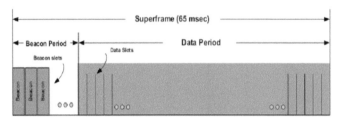

Figure 3.3 Wimedia MAC Superframe structure.

The total time length of the Superframe is 65,536 microseconds, and it is divided into 256 media allocation slots (MAS), and each of these MAS is 256 microseconds long. The three major subdivisions of the Superframe are:

- Beacon period that starts at the beacon period start time (BPST);
- Beacon slots;
- The media access slots.

If a device wishes to participate in the network, it is required to synchronize its timing to coincide with the BPST, so that the Superframe timing is shared among all network members. After the establishment of synchronization, a new device is required to listen to the existing Superframe contents as the current network status is represented in the beacon slots. The beacon portion of the Superframe acts as a kind of communal bulletin board: the Superframe not only stores information about the node itself but also keeps the information about the neighbours that it can hear. In doing so, any individual node is able to create a map of devices up to two hops away, which can be used to coordinate activities.

3.3 Node Discovery Scenario

We designed our mechanism for scenarios where one device wants to discover an existing cooperative cluster that lies in the device's area. The proposed network scenario is shown in Figure 3.4 and involves MTs that can communicate through a long range network (e.g. WiMAX, UMTS) and a short range network that makes use of super-frames (e.g. UWB, Bluetooth). We focus on just one pair of technologies and consider WiMAX and WiMedia UWB, but the approach has general validity. The scenario covers cluster discovery and cooperation instantiation for a mobile terminal that is in search of a network or a device. The cluster is based on the UWB and works in a centralized setup but is also applicable in a decentralized way with the neighbor nodes. Our

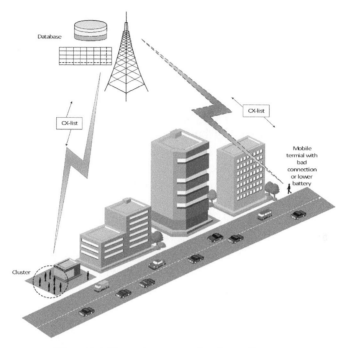

Figure 3.4 Network structure for cluster discovery.

approach uses the context information about the cluster e.g. battery levels, and willingness to cooperate. Furthermore, inside the cluster we have defined two groups of member devices: the first group consists of devices that are considering switching to the hibernation state to save their energy, and the second group includes more "stable" devices that are suitable for cooperation. The traditional approach to device discovery relies on a long "scan" period, where the MT scans different channels while staying in "listen" state, looking for beacons. In the proposed approach, on the other hand, the MT uses context information to facilitate the discovery of the cluster.

3.4 Proposed Approach

For node discovery and cooperation in the above scenarios we propose a mechanism that facilitates the short range searching, to allow a device to find its nearby cluster node and choose the best for communication and cooperation.

Table 3.1 Willingness IE format.

Octet 1	1	1
Willingness	Length	Element ID

3.4.1 Proposed Beacon Approach

This approach proposes to modify the superframe of the short-range communication protocol. Instead of building the communication of beacons on contention access, we propose inserting a superframe map in the beacon, bearing information about which part of the beacon slot was reserved by remote terminals. This way, using context information exchanged via long-range RATs, devices are able to know in advance when they have to send their beacons to the cluster to complete the rendezvous. The rest of the beacon slots will be used via PCA, to welcome devices that are not exchanging context information via long-range RATs. This way, the insertion of devices that use context information in the cluster is faster, while permitting other MTs into the cluster via the traditional beacon exchange. We propose that an MT, connected via long-range RAT to the network, exchanges information with the cluster head of a cluster to inform the cluster head that the MT wishes to access the cluster. The cluster-head will add a field to the superframe, at the beginning of the first beacon, to inform all the devices that some of the beacon slots are reserved for devices that are using context information. Then, the clusterhead sends a number to the MT via the long-range RAT, indicating which beacon frame was reserved for it. As soon as MT detects the start of a super-frame, it can wait for its own reserved beacon frame and immediately start sending its own beacon.

3.4.2 Willingness Representation in Beacon

We propose that the mobile devices send their battery level and their willingness to cooperate to the cluster head, which exploits these IEs to take energy-efficient decisions. A first example of IE for willingness is shown in Table 3.1 and the format of the field containing the Willingness status is reported in Table 3.2. The willingness for cooperation of the device is represented by two bits as:

$00 \rightarrow$ not willing to cooperate.
$01 \rightarrow$ willing to cooperate and to be relayed.
$10 \rightarrow$ willing to relay.
$11 \rightarrow$ willing to cooperate both as relay and relayed.

Table 3.2 Format for willingness status field.

b7-b2	b0-b1
Reserved	Status

Table 3.3 Battery level IE format.

Octet 1	1	1
Battery level	Length	Element ID

3.4.3 Energy Representation in Beacon for Cooperative communication

This section discusses our way to integrate energy encoding with willingness encoding in a beacon frame. This representation is alternative to the one that was proposed in Section 3.4.2, which only provides the Boolean "willingness to cooperate" of a device. In the previous approach, we have different fields in the transmitted packets, containing the different levels of willingness to relay and to be relayed, and the energy level of the unit. The energy level will be either a percentage of the maximum energy level, or an estimation of the unit's lifetime if its energy consumption will be kept constant, or a prediction about the lifetime the unit will enjoy with its energy consumption profile. Anyway, in the case of node discovery for energy saving, the principle of knowing the energy level of a unit and its willingness to cooperate is to evaluate if it should act like a relay or a source. Thus, for the purpose of saving energy we devised a way to encode all this information in a small number of bits that are used by a MT to map its energy level, depending on its willingness to relay or to be relayed as shown in Figure 3.5.

The MT can prefer to be a relay, to be relayed, or can be willing to be both a relay or relayed. The MT will bias the energy information it sends to act more often in the way it prefers, as represented in rows a, b and c in Figure 3.5. Numerical values for the implementation of the proposed "biasing system" for the battery level are shown in Tables 3.3 and 3.4.

Let us consider three different MTs, that have got three different goals, and that are described graphically in Figure 3.5 in rows a, b and c. If a unit wants to relay (see Figure 3.5a), the ranges of energy that will be mapped to the lowest values will be smaller, to ensure that the MT is considered as a

Table 3.4 Battery level field format.

b7-b4	B3-b0
Reserved	Battery Status

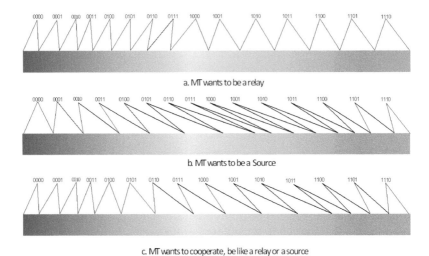

a. MT wants to be a relay

b. MT wants to be a Source

c. MT wants to cooperate, be like a relay or a source

Figure 3.5   MT's energy and role mapping in cooperation.

candidate for relaying. On the other hand, if a MT wants to be relayed (see Figure 3.5b), the energy levels mapped to the lowest values will be larger. The result is that, when the system will compare the two MTs' energy levels, it will assign the role of relay to the one that encoded its energy level to a higher number, and of source to the other. Figure 3.5c shows a more balanced energy coding, where a MT wants to cooperate and is willing to act both as a relay and a source. The battery levels representation is shown in Table 3.5.

3.4.4 Context Information Management at a MAC Layer

Context awareness for energy saving is at the core of the optimization in many research projects, to enable a network element to take decisions and to instantiate a particular energy saving approach. In our work, we particularly consider that a MT requires information on a number of features, settings and preferences/policies of the networks or other nodes. Figure 3.6 presents our context based management MAC layer for the MT. The figure shows how the context list is constructed and how the information is extracted from the context list. When the beacon frames are received at the MAC layer the CxEF (context extraction function) extracts the different bits from the beacon and matches them to the defined values, and finally forwards the extracted information. The extracted information can be further utilized in different actions of the mobile terminal, which are not under the scope of this

Table 3.5 Representation of battery levels.

Bits	Basic representation	Relay	Source	Both
0000	6%	4%	8%	4%
0001	13%	9%	17%	9%
0010	20%	14%	26%	14%
0011	26%	18%	34%	19%
0100	33%	23%	43%	26%
0101	40%	28%	52%	35%
0110	46%	32%	60%	45%
0111	53%	39%	67%	54%
1000	60%	48%	72%	63%
1001	66%	56%	76%	71%
1010	73%	65%	81%	80%
1011	80%	74%	86%	85%
1100	86%	82%	90%	90%
1101	93%	91%	95%	95%
1110	100%	100%	100%	100%
1111	not willing	not willing	not willing	not willing

chapter. For instance, the battery level information could be used in energy efficient routing or cooperation. On the other hand, to construct the context list and embed the information in the beacon frame, the CxEF extracts both the dynamic and static context information of the terminal and inserts it in the beacon according to the requirements. The CxEF works at the UWB's MAC layer.

3.5 Simulation Setup and Results

The proposed node discovery mechanism has been simulated using Omnet++ 4.0 (an open source simulator) [20]. The simulation consists of a cluster of 4 nodes and a separate node, which is willing to discover nodes/cluster for cooperation. The discovering node moves with different speeds and mobility patterns and searches for cooperating nodes/cluster by using both context-based discovery and normal periodical scanning. The simulation setup is shown in Table 3.6.

In the proposed scenario, if MT-A (the discovering node) does not find any cooperative node or cluster, MT-A has to continue both its connection with the long range technology, and its scanning process; hence its battery will drain out quickly. Figures 3.7 and 3.8 provide a graphical representation

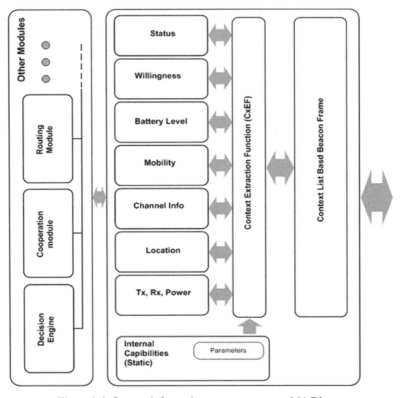

Figure 3.6 Context information management at a MAC layer.

of why we need short range technologies for energy saving, and they also validate our energy saving mechanism that uses WiMAX as facilitating technology in node discovery process of UWB nodes. We have generated a series of traffic bursts with different payload size and checked energy consumption in UWB and WiMAX. Figure 3.7 shows that long range communication is expensive in terms of energy consumption. The energy consumption of MTs also depends on the data rates, payload size and the distance between the communicating devices. Since UWB offers high data rates, the transmission time is much smaller than in the case of WiMAX, and the results shown in Figure 3.8 confirm that the transmission time is proportional to the data payload size.

To facilitate the task of looking for a cooperative node or cluster, the cluster heads gather information about all cluster member nodes into a context list, and send the list to the WiMax base station. The context information

Table 3.6 Simulation setup.

Parameters	Value
Simulation Area	500 m × 500 m
Short range Technology	UWB
Long range Technology	WiMAX
Battery module	Omnet++ 4.0 Simple battery
Nodes in the cluster	10
Mobility Models	Constant Speed, Linear Mobility, Rectangular (Defined areas)
Node speed	2 m/s
Node update interval	1 s
Number of nodes	10–100
Cx-list size	32, 64, 128, 256, 512 bytes
Initial node Battery	7 mAh
Voltage	3.3 V
Maximum transmission power (UWB)	1 Mw
Backoff time	0.0003 s
SIFS	0.00019 s
Time from RX to TX mode	0.00018 s
Time from RX to Sleep mode	0.000031 s
Time TX to RX mode	0.00012 s
Time TX to Sleep mode	0.000032 s
Time Sleep to RX mode	0.000103 s
Time Sleep to TX mode	0.000203 s

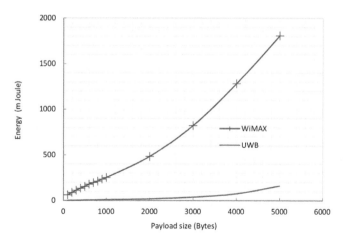

Figure 3.7 Energy consumption of different payload sizes over UWB and WiMAX.

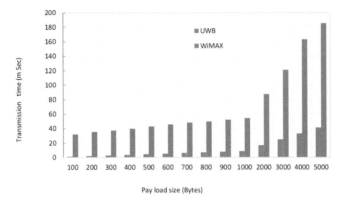

Figure 3.8 Transmission time UWB and WiMax with data payload size.

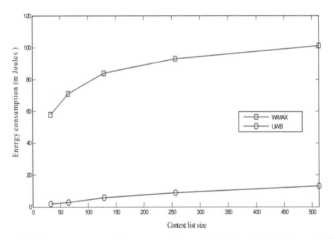

Figure 3.9 Energy consumption comparisons of Cx-list via WiMax and UWB.

is transmitted via WiMax connection and received by the mobile device. Figure 3.9 shows the energy consumption of context lists with different sizes over UWB and WiMax. We have varied the size of Cx-list from 32 bytes up to 512 bytes to check the energy consumption variation in order to validate our mechanism. The results show that the energy consumption for different sizes of Cx-list of WiMAX is higher than UWB but negligible compared to the overall gain of the mechanism.

The time to discover the cluster varies with different mobility models, and this time impacts the energy savings, since in the context based discovery the MT waits all this time before activating the short range interface, hence it can

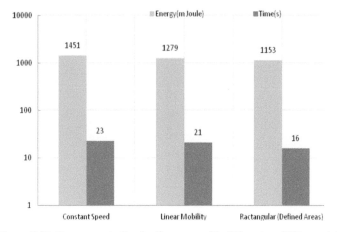

Figure 3.10 Energy cost of node discovery with different mobility models.

keep the short range interface off and save energy. The energy consumption of node discovery with different mobility models is shown in Figure 3.10 We have used Constant speed, and mobility over one-dimensional (Linear mobility) and two-dimensional (rectangular mobility) areas. The node's transmission power has been kept constant to 1mW and the speed to 2 m/s in all simulation runs. Figure 3.11 shows the energy consumption of a node when it scans blindly (without context information) to find a cooperative cluster, and when it performs context-based scanning. A number of experiments have been performed to compute the average time and consumed energy for the node to get within the range of the short range technology of the cooperating cluster. Despite using a MAC that has reasonable duty cycle of sleep and active mode, the energy consumption of the UWB interface of the discovery node is high. Figure 3.11 shows that the context based discovery process saves a considerable amount of energy.

With the traditional scanning process, the MAC layer periodically switches to sleep and active modes. By active mode we mean both sending and listening states while the sleep mode consider both idle and sleep modes. In absence of prior knowledge of nearby nodes or clusters, the node scanning process is quite expensive in terms of energy. On the other hand, our node uses context based scanning to become active and start communication just-in-time when it reaches the coverage area of the cluster. As shown in Figure 3.12, terminals can save 50% of their energy in the short range discovery process compared to the normal discovery process.

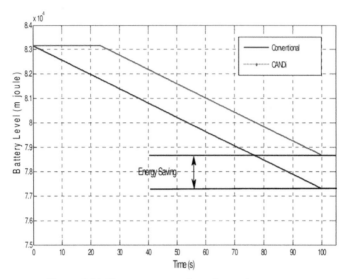

Figure 3.11 Energy consumption of scanning process.

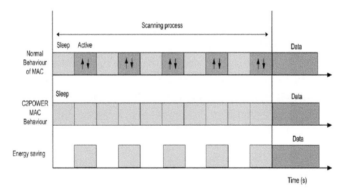

Figure 3.12 Energy consumption of scanning process.

3.6 Conclusions

Node discovery is the initial step to initiate cooperative communication. In this chapter we have presented a new mechanism for context based node discovery, where long range technology (WiMax) supports short range technology (UWB) in the discovery process. To make use of the available beacon IEs of UWB, we have proposed novel IEs representing the energy level of batteries and nodes willingness to cooperate, which further contribute to the cooperative cluster building. We have shown how a context list is constructed and how the information is extracted from the context list at MAC layer for further utilization in other operations of the terminal. By means of simulations, we showed that our context based discovery mechanism contributes to energy saving in the scanning process, and prolongs the MTs overall life time. As node and neighbor discovery can drain out a significant battery of MTs due to excessive scanning, a more comprehensive framework for node discovery is the future aim of this work. Context information, of both terminal and network, is utilized to facilitate MT discovery of other terminals and/or clusters for energy efficiency. A database connected at the localized base station side can provide all the location information to the central base station. The refined context information from the central database then guides the MT to discover the most energy efficient node for communication. The flow of context information to and from central database can cause some signaling overhead; this overhead will be analyzed via analytical model and simulation results.

Acknowledgements

The research leading to these results has received funding from the European Community's Seventh Framework Programme (FP7/2007-2013) under grant agreement No. 248577 (C2POWER). Muhammad Alam, PhD student with the MAP-I, would like to acknowledge the support of the grant of the Fundacão para a Ciência e a Tecnologia (FCT, Portugal), with reference number SFRH/BD/ 67027/2009.

References

[1] TNS online report, two-day battery life tops wish list for future all-in-one phonedevice, http://www.tns.lv/?lang=en&fullarticle=true&category=showuid&id=2288/.
[2] C2 power project, http://www.ict-c2power.eu/.
[3] Greent project, http://greent.av.it.pt/.
[4] M. A. Ameen, S. M. R. Islam, and K. Kwak, Energy saving mechanisms for mac protocols in wireless sensor networks, *International Journal of Distributed Sensor Networks*, vol. 2010, October 2010.
[5] M. S. I. M. Zin and M. Hope, A review of UWB MAC protocols, in *Proceedings of the 2010 Sixth Advanced International Conference on Telecommunications*, Ser. AICT '10, pp. 526–534. IEEE Computer Society, Washington, DC, 2010. [Online] Available: http://dx.doi.org/10.1109/AICT.2010.101.
[6] Wimedia standard, http://www.wimedia.org/en/index.asp/.

[7] J. Reed, R. M. Buerhrer, and D. Mckinstry, Introduction to UWB: Impulse radio for radar and wireless communications. GM briefing, August 2002, http://sss-mag.com/pdf/uwbcars.pdf/.

[8] Y. M. Kim. Ultra wide band technology and applications. Nest group , July 2003.

[9] K.-H. Liu, X. Ling, X. Shen, and J. Mark, Performance analysis of prioritized mac in uwb wpan with bursty multimedia traffic, *IEEE Transactions on Vehicular Technology*, vol. 57, no. 4, pp. 2462 –2473, July 2008.

[10] D. T. C. Wong, F. Chin, M. Shajan, and Y. Chew, Performance analysis of saturated throughput of PCA in the presence of hard DRPS in Wimedia MAC, in *IEEE Wireless Communications and Networking Conference, (WCNC 2007)*, pp. 423 –429, March 2007.

[11] K.-H. Liu, X. Shen, R. Zhang, and L. Cai, Performance analysis of distributed reservation protocol for UWB-based WPAN, *IEEE Transactions on Vehicular Technology*, vol. 58, no. 2, pp. 902–913, February 2009.

[12] W. Lu, A. Lo, and I. Niemegeers, On the dynamics and self-organization of personal networks, November 2004.

[13] Bluetooth, http://www.bluetooth.com/.

[14] M. J. McGlynn and S. A. Borbash, Birthday protocols for low energy deployment and flexible neighbor discovery in ad hoc wireless networks, in *Proceedings of the 2nd ACM international symposium on Mobile Ad Hoc Networking & Computing*, ser. MobiHoc '01, pp. 137–145. ACM, New York, 2001. [Online] Available: `http://doi.acm.org/10.1145/501431.501435`.

[15] R. Garces and J. Garcia-Luna-Aceves, Collision avoidance and resolution multiple access for multichannel wireless networks, in *Proceedings of INFOCOM 2000, Nineteenth Annual Joint Conference of the IEEE Computer and Communications Societies*, vol. 2, pp. 595–602. IEEE, New York, 2000.

[16] K. Prasad, *Principles of Digital Communication Systems and Computer Networks*. Charles River Media, 2003.

[17] J. H. Reed, *An Introduction to Ultra Wideband Communication Systems*. Prentice Hall, 2005.

[18] F. Baccelli, B. Blaszczyszyn, and P. Muhlethaler, An ALOHA protocol for multihop mobile wireless networks, *IEEE Transactions on Information Theory*, vol. 52, no. 2, pp. 421–436, February 2006.

[19] S. Wood and R. Aiello, *Essentials of UWB*. Cambridge University Press, 2008.

[20] Omnet++, http://www.omnetpp.org/.

4

Throughput Fairness Analysis of Reservation Protocols for WiMedia MAC

Muhammad Alam[1,2], Michele Albano[3], Ayman Radwan[1] and Jonathan Rodriguez[1]

[1]*Instituto de Telecomunições, Aveiro, Portugal*
[2]*Universidade de Aveiro, Aveiro, Portugal*
[3]*CISTER/ISEP Polytechnic Institute of Porto, Portugal*
e-mail: alam@av.it.pt

Abstract

Medium Access Control (MAC) layer protocols play an imperative role to handle the challenging QoS provisioning issue by efficiently controlling the accessing and utilizing of wireless channels. WiMedia standardized a fully distributed high data rate communication using Ultra Wideband (UWB) for wireless personal area networks (WPANs). UWB offers a number of advantages over other short range technologies, e.g. high data rates, low power and precise positioning, which makes it more suitable for the WPAN. The WiMedia device accesses the channel via a superframe and the Medium Access Slots (MASs) in superframe can be reserved via Distributed Reservation Protocol (DRP) or accessed via Prioritized Contention Access (PCA). This chapter presents the analysis of the WiMedia reservation protocols. By means of simulations and analytical results we show how the QoS provisioning issue is handled by the MAC protocols and how fairness is provided to users in terms of throughput and delay, shortening the time they spend trying to access the wireless medium. In turn, the net effect contributes to energy saving for data communication at the nodes, and adds to the Quality of Experience (QoE) of the user.

Shahid Mumtaz and Jonathan Rodriguez (Eds.), Green Communication in 4G Wireless Systems, 53–68.

Keywords: Medium Access Control, Quality of Service, WiMedia, Ultra Wideband.

4.1 Introduction

The protocols of the Medium access control (MAC) layer play an imperative role to handle challenging QoS provisioning issues by efficiently controlling how the terminals access and utilize the wireless channels. In July 2008, WiMedia Alliance and European Computer Manufacturers Association (ECMA) International announced a joint development relationship. Both partners agreed upon the development of specific standards, which will be conducted by joint work of the engineers of the member companies. For high data rates wireless personal area networks (WPANs), the ECMA standard [1] defined Physical and MAC layers that offer a number of policies and control mechanisms to ensure the quality of service (QoS) provisioning. Ultra Wideband (UWB) offers a number of advantages over other short range technologies e.g. high data rates, low power consumption and precise positioning capabilities, which make it more suitable for the WPAN. UWB supports data rates ranging from 53.3 to 480 Mbps over distances up to 10 meters [2].

MAC protocols can be contention free and contention based, and some proposed ones operate under a combination of both paradigms. Contention based protocols waste large quantities of energy, since the nodes have to spend long periods of time with their radio on to estimate if a channel is available. Contention free protocols require more challenging mechanisms to handle access to the channel by competing nodes with different traffic types. WPAN topology designs can be centralized or decentralized. IEEE 802.15.3 [3] is an example of a centralized approach where the devices form a cluster called Piconet and a central device, called Piconet Coordinator (PNC), manages all the members of the piconet. The PNC has the knowledge of the member nodes and assigns the available Medium Access Slots (MASs) following the Time Division Multiple Access (TDMA) approach. But the centralized approach has a number of limitations, e.g. QoS provisioning, scalability of the network and the PNC represents a single point of failure since, if it fails, the whole network goes down. Furthermore, the connectivity of multiple Piconets is costly in terms of inter-piconet interference and leads to degraded performance [4]. To address these problems a distributed MAC, which is based on multiband orthogonal frequency-division multiplexing (MB-OFDM), has been proposed by WiMedia Alliance [5] and ECMA [1]. Like IEEE 802.15.3, WiMedia MAC is based on slotted superframes. Each

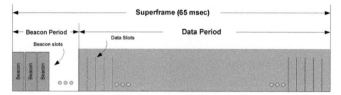

Figure 4.1 Superframe.

node tries to get or reserve slots in the superframe, which is managed by reservation protocols defined in the WiMedia standard. In this chapter we propose three approaches to using the reservation process of WiMedia MAC for efficient and fair access to the wireless medium. By means of simulation, we show that the medium access is more efficient than with the current standard mechanisms, which in turn shortens the total time spent in the communication process and improves the energy efficiency of the communication technology.

4.2 Overview of the UWB MAC Layer

This section presents an overview of the Wimedia UWB MAC.

4.2.1 Superframe

The WiMedia MAC channel time is divided into Superframes. The total duration of the Superframe is 65,000 s and is composed of 256 MASs with duration of 256 μs each [1]. The Superframe is actually divided into two main parts, a Beacon Period (BP) and a Data Transfer Period (DTP), as shown in Figure 4.1. In BP each user transmits its own beacon frames which contain a number of Information Elements (IEs). These IEs have timing, control and synchronization information of users to access the channel in a fully distributed manner. The beacon frames represent both the users' information, and their views of the network, which help the incoming users to identify empty beacon slots, occupy them and transmit their own beacons. The beacon frames are also used to reserve MASs in the DTP.

4.2.2 Distributed Reservation Protocol (DRP)

The usage of MASs in the DTP is reserved, modified or released via DRP, or accessed via Prioritized Contention Access (PCA). DRP is used to reserve the MASs mostly for isochronous traffic, or for nodes that need guaranteed ac-

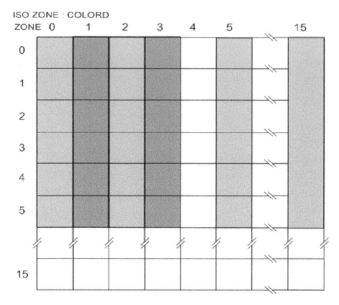

Figure 4.2 Superframe's two-dimensional view.

cess to the medium. On the other hand, PCA provides differentiated channel access to the medium similar to IEEE 802.11e. When a node wants to reserve MASs for data transmission or reception, it negotiates with its neighbors via DRP and reserves a set of MASs. The DRP frame contains a number of IEs representing different pieces of information. The DRP contains the control IEs, which show owner, status of reservation, reason codes, reservation types and some more information about reservation conflicts [1]. The device that wants to start the reservation process is called the owner and the device that receives the information for reservation is referred to as the target. The type of reservation can be Hard, Soft, PCA, Private or Alien BP. The reservation status indicates the status of the DRP negotiation process, which shows whether a reservation is under negotiation, in conflict or established. The Reason Code is used by a reservation target and it shows whether a DRP reservation request was successful or not.

A DRP IE contains one or more DRP Allocation fields which are encoded using a zone structure. The zone is composed by 16 zones numbered from 0 to 15 starting from the BP, and are further divided into isozones that are separated by fixed intervals. In this two-dimensional structure of the WiMedia superframe each column of the superframe matrix is called an allocation

Table 4.1 Access categories (ACs) in PCA.

Priority	User Priority	AC
Lowest	1	Background
↓	2	Best Effort
↓	3	Video
Highest	4	Voice

zone (see Figure 4.2). In the reservation allocation IE each node includes a Zone bitmap and a MAS bitmap, where Zone bitmap identifies the zones that contain reserved MASs and the MAS bitmap specifies which MASs in the zones identified by the Zone Bitmap field are part of the reservation. The reservation of MASs in the zones follows certain rules to ensure fairness; details are available in [1].

4.2.3 Prioritized Contention Access (PCA) Protocol

PCA is based on carrier-sense multiple access with collision avoidance (CSMA/CA) and employs different contention parameters in order to support both non-real-time and real-time data transfers, and to contribute to network scalability. In PCA, four access categories (ACs) of traffic are defined, which are called voice, video, best effort, and back-ground [1], as shown in Table 4.1.

The PCA procedures are applied by any devices for each AC to obtain a transmission opportunity (TXOP) for the frames belonging to that AC using the PCA parameters associated with that AC. Each time a device has a frame to transmit it will first sense the channel and occupy a free channel to start communication with the target device. If the channel is busy, it goes to back-off state and sets a backoff counter value, ranged from minimum contention window (mCWmin) to maximum contention window (mCWmax). Therefore, for higher priority ACs, the value of CW must be lower so that the higher priority node spends less time in the backoff state. Figure 4.3 shows an example for access ACs with different priorities. The priorities are represented by the length of arbitrary inter-frame space (AIFS). Voice is given high priority with a shorter AIFS, followed by video and then by best effort traffic. A device shall wait for the medium to become idle for AIFS[AC] seconds before obtaining a TXOP or starting/resuming decrementing the backoff counter for the AC. Each device sets the value of CW[AC] to a value in range of minCW and maxCW after invoking a backoff for the AC.

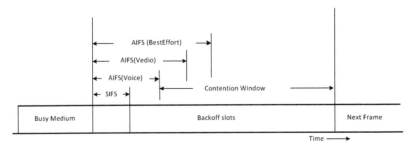

Figure 4.3 PCA for different ACs.

Table 4.2 WiMedia's UWB data rates.

Data Rate Mb/s	Modulation	Coding Rate
53.3	QPSK	1/3
80	QPSK	1/2
106.7	QPSK	1/3
160	QPSK	1/2
200	QPSK	5/8
320	DCM	1/2
400	DCM	5/8
480	DCM	3/4

4.2.4 Throughput of WiMedia MAC

The ECMA standard specifies that WiMedia uses the ultra-wideband (UWB) physical layer, which transmits over the unlicensed 3100-10600 MHz frequency band. The data rates supported by this standard are 53.3, 80, 106.7, 160, 200, 320, 400, and 480 Mb/s, and the support for transmitting and receiving data rates of 43.3, 106.7, and 200 Mb/s is mandatory [1]. The data rates are dependent on modulation and coding rates, as shown in Table 4.2.

4.3 Reservation Fairness Using Gini Index

The Gini index represents a measure of statistical dispersion developed by the Italian statistician and sociologist Corrado Gini [6]. The Gini coefficient is a measure of the inequality of a distribution, with the value of 0 expressing perfect equality and the value of 1 expressing maximal inequality. We applied the Gini index to the fair distributed resource distribution problem of WiMedia MAC protocol. We used Gini index to calculate the throughput fairness by the following formula. Let l and k be two users having observed throughputs, then the Gini index for throughput fairness is calculate as shown in equation (4.1):

$$G = \frac{\sum_{k=1}^{n} \sum_{l=1}^{n} |\Theta_{av\_k} - \Theta_{av\_l}|}{2n^2 \bar{\Theta}_{av\_R}} \qquad (4.1)$$

where Θ is the observed Average Throughput value of users k and l; n is the number of average throughput observed; $\bar{\Theta}_{av\_R}$ is the average throughput of all users placed within the coverage R.

4.4 Related Work

DRP analysis, with different parameters and approaches, has been shown in previous works, both by analytical models and simulations. In [7] the MAC layer is modeled as a queuing system by employing Markovian Arrival Process (MAP) and PHase type distribution (PH), and the throughput is analyzed with respect to traffic load. In [8] the authors presented a study on the delay performance of DRP channel access for Multi-band OFDM Alliance (MBOA) UWB MAC by producing a bi-dimensional Markov chain model, where one dimension represents queue size distribution and the other is for allocated slots. UWB-based WPAN, where physical layer adaptive modulation and coding is coupled with the DRP and the delayed acknowledgement schemes at the link layer, has been studied in [9]. A work based on traffic type and priority reservations is presented in [10], while in [11] the blocking probabilities of reservations is considered. Another analytical model for DRP based on a shadowing effect that varies the channel condition has been studied in [12]. The authors consider both hard and soft DRP, and have proposed a channel model to describe the dynamic behavior of the UWB shadowing channel at the packet level. On a first-come first-served basis, a contention-free distributed protocol has been analyzed for delay and throughput fairness in [13]. A study on the delay performance of DRP under different reservation patterns under the dynamics of UWB shadowing channel has been presented in [14]. The authors have modeled the system as a discrete-time single server queue with vacation represented by the quasi-birth and death process to analyze the performance of DRP. In [15] the authors have presented a work on the performance analysis of PCA protocol considering the bursty nature of multimedia traffic. By means of simulation and analytical results, it has been shown that the effect of the traffic differentiation mechanism in PCA is magnified when the inter-arrivals are highly bursty and correlated. To derive the MAC service time, discrete Markov modeling approach has been applied in [16–21] and its mean value analysis is provided in [22, 23].

In this chapter we show how the incoming nodes with different traffic types affect the reservation process of an already connected network. We have used hard, soft and PCA reservation types to allocate the MASs to nodes with mixed traffic (voice, video, best effort). The incoming nodes use first come first served scheduling to access the channel.

4.5 Proposed Approaches for Analysis

In this section we present our approaches to analyze the WiMedia's MAC reservation protocols. The DRP plays an important role to guarantee the QoS of isochronous traffic. The devices use IE in the beacon period to reserve the MASs in the Superframe for successive data transmission. Since the MASs are allocated by DRP without prior knowledge of the traffic, to ensure overall traffic load balance or fair distribution of the MASs among the nodes without a centralized controller, we propose approaches to test the performance of the reservation based on the traffic types of incoming mobile nodes into the system. The nodes can use the DRP as a selfish approach to occupy the MASs for extended duration ignoring the waiting nodes or less priority traffic nodes such as best effort traffic. This selfish approach not only restricts the size of the network, but also degrades the overall performance of the network. Moreover, when a node is waiting to access the network, it is still spending energy in radio-related tasks, hence its performance in terms of bits/J becomes worse. Furthermore, the standard [1] defines a number of DRP reservation methods e.g Hard, soft, private, PCA, etc. so MAS allocation needs to be carefully handled during the beacon period and proper MAS access mechanisms should be used.

In our approaches we have divided the Superframe into parts for DRP and PCA users to check the effect on throughput and to keep a balance between the two resource allocation mechanisms with mixed traffic.

4.5.1 Analysis Approach 1

The reservation process becomes more challenging when the nodes are mobile, and detached nodes and/or incoming nodes want to join the network. In this first approach, we divide the superframe into two equally sized parts, DRP part and PCA part, as shown in Figure 4.4. For isochronous traffic (voice and video) nodes reserve the MASs based on hard DRP; once they

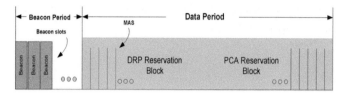

Figure 4.4 Final Superframe of Approach 1.

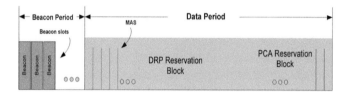

Figure 4.5 Final Superframe of Approach 2.

occupy specific MASs, they have to release them as soon as their data transfer ends. Furthermore, Isochronous traffic nodes cannot reserve the PCA MASs and PCA nodes cannot occupy the DRP MASs. The incoming nodes are not allowed to use the relinquish request IE once a MAS is reserved by Hard reservation. The purpose of this approach is to check whether we can provide a balance to the MAS allocation by DRP and PCA and also to check the effect on PCA-based reservation in the presence of DRP Hard reservation for both types of traffic.

4.5.2 Analysis Approach 2

In the second approach, based on the priority requirements for voice and video traffic, we reserve more MASs for the users who need more guarantees in QoS. The final superframe is shown in Figure 4.5. The nodes can reserve MASs by both Hard and Soft reservation. The Isochronous traffic nodes cannot reserve the MASs reserved by PCA and the other way around is valid for PCA reservation. Incoming nodes can use the relinquish request IE to inform the neighbors to release the MASs occupied by Soft reservation or PCA. The PCA based reservation nodes will go to backoff once they find that the PCA portion of MASs is saturated.

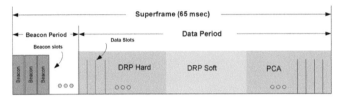

Figure 4.6 Final Superframe of Approach 3.

4.5.3 Analysis Approach 3

In this approach we use dynamic reservation that is based on the traffic type and does not specify or fix the MASs for DRP and PCA reservations.

The final superframe of the proposed analysis approach is shown in Figure 4.6 in which nodes can reserve MASs by both Hard and Soft reservations, and can also benefit from PCA. The Isochronous traffic nodes can reserve the PCA MASs if there are any available, by following the rules of ECMA standard. If a node with isochronous traffic wants to enter the system, it will first check free MASs to reserve, then it will go to backoff state and it will wait for PCA MASs to be free, and not finding any for a long time it will request for releasing the MASs reserved by Soft reservations. Nodes that are connected and those which are getting into the system both use the DRP service primitive parameters, e.g. desired bandwidth, available bandwidth and minimum bandwidth, to know about the number of MASs needed for their data transmission. Thereafter nodes use these parameters and estimate how much bandwidth they need, and how much is available in the system.

4.6 Simulation Scenario and Setup

This section presents our scenario for simulations, which is represented graphically in Figure 4.7. For analyzing the proposed approaches, we consider random arrival of nodes with mixed traffic requirements into the system. The above scenario is simulated using Omnet++4.1 [24] , an open source simulator. Different results are generated for the analysis of DRP and PCA under the rules of proposed approaches. The connected network is composed of six nodes and the number of incoming nodes varies which will affect the existing reservations in the superframe. The nodes joining the network have a constant speed of 0.5 m/s. All the nodes have a maximum transmission power of 1 mW. The data rate is 480 Mbps and the traffic type is mixed (voice,

Connected Network

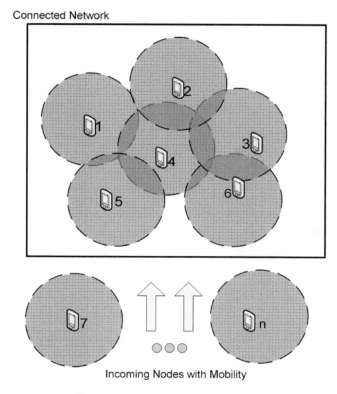

Incoming Nodes with Mobility

Figure 4.7 Simulated network scenario.

video, best effort). The scheduling of the allocation process is maintained on first-come first-served basis. The duration of each time slot is set to 256 μs.

4.7 Simulation Results

Initially, when the system is not saturated, the incoming nodes with both types of reservation - PCA and DRP - and under all the three approaches can be accommodated into the network. But with the increase of the arrival rate of new nodes the probability of successfully reserving the MASs to transmit data becomes lower, especially for nodes using DRP. Figure 4.8 shows the amount of allocated MASs to DRP and PCA reservation for the proposed approaches. In the case of Approach 1, less MASs are allocated to DRP nodes because of the Hard reservation, whereas MAS allocation is prevalent for PCA traffic

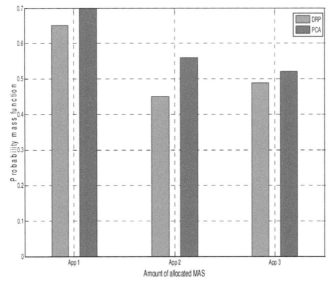

Figure 4.8 Allocation of MASs with DRP and PCA for the three proposed approaches.

nodes because a large portion of superframe is assumed to be utilized by best effort traffic nodes.

Approach 2 provides more flexibility to the isochronous traffic nodes by using both Hard and Soft reservation, but still cannot make optimized use of the free MASs maintained for incoming PCA nodes. In Approach 3, the incoming nodes scan the reserved MASs and adopt the policy to reserve the MASs in a more flexible way, thus providing a more balanced provisioning of MASs allocation to both isochronous and asynchronous traffic nodes.

Figure 4.9 shows the throughput of the system with the arrival of DRP and PCA traffic nodes. Initially the number of nodes in the system is lower and there are free MASs available for incoming reservations. The proposed approaches contribute differently to the system overall throughput because of the different reservation types and rules. Approach 3 shows a high throughput because of Soft reservation and PCA's MASs occupation. It reaches its best value for 20 nodes, and then remains constant because of saturation. The throughput of Approaches 1 and 2 is low because of the fixed MASs for traffic types, and because they depend on the traffic of nodes joining the system. Therefore, the throughput of the system depends on the incoming node traffic type, rate and also the followed approach. In general, a higher throughput produces derived benefits, since it corresponds to a communication process

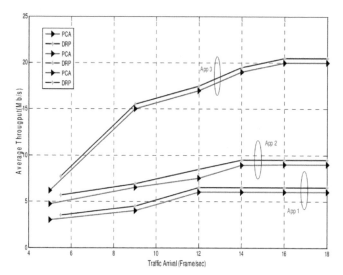

Figure 4.9 Throughput of the DRP and PCA with proposed approaches.

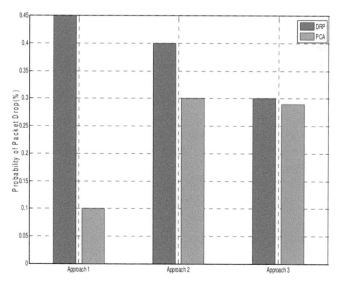

Figure 4.10 Packet drop of the DRP and PCA with proposed approaches.

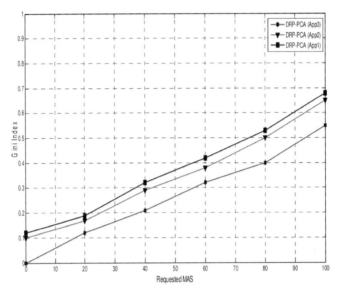

Figure 4.11   Gini index of allocated MASs to the incoming DRP and PCA nodes.

that completes in a shorter time, hence leading to switching off the radio after a shorter timespan, with the net result of a higher bit/J efficiency.

The number of packets in the system and their successful delivery depend on the flexibility of approach, rate of arrival, rules and policies applied to incoming traffic classes, and finally on the buffer size of the nodes. In case of usage of Hard reservation, and when keeping the superframe reservation portions fixed for both traffic type, the system ends up having high packet drops. On the other hand, providing more flexibility to the reservation, by utilizing the available MASs and keeping a balance between PCA and DRP Hard and Soft reservation, the number of packet drop decreases. The packet drop of the incoming nodes for the three approaches is shown in Figure 4.10.

The Gini index is calculated using the throughput of 30 nodes, according to equation (4.1). As shown in Figure 4.11, Approach 3 provides better fairness in term of throughput for all the nodes, as MASs can be relinquished by utilizing Soft reservation and also by estimating the number of available MASs for a specific incoming node.

Acknowledgements

The research leading to these results has received funding from the European Community's Seventh Framework Programme (FP7/2007-2013) under grant agreement No. 248577 (C2POWER). Muhammad Alam, PhD student with the MAP-I, would like to acknowledge the support of the grant of the Fundacão para a Ciência e a Tecnologia (FCT, Portugal), with reference number SFRH/BD/ 67027/2009.

References

[1] ECMA 3rd edition, www.ecma-international.org/publications/standards/Ecma-368.htm/, December 2008.

[2] W. P. Siriwongpairat and K. J. R. Liu, *Ultra-Wideband Communications Systems*, 1st ed., John Wiley and Sons, New Jersey, 2007.

[3] IEEE 802.15.3, Wireless medium access control and physical layer specification for high rate wireless personal area networks, 2003.

[4] L. Cai, X. Shen, and J. Mark, Efficient MAC protocol for ultra-wideband networks, *IEEE Communications Magazine*, vol. 47, no. 6, pp. 179–185, June 2009.

[5] WiMedia Alliance. WiMedia MAC Release Spec. 1.01. Distributed medium access control for wireless networks. [Online] http://www.wimedia.org/en/index.asp./, December 2006.

[6] Gini index, http://en.wikipedia.org/wiki/Gini_coefficient/.

[7] N. Arianpoo, Y. Lin, V. Wong, and A. Alfa, Analysis of distributed reservation protocol for UWB-based WPANs with ECMA-368 MAC, in *Proceedings of IEEE Wireless Communications and Networking Conference (WCNC 2008)*, 31 March–3 April 2008, pp. 1553–1558. IEEE, 2008.

[8] H. Wu, Y. Xia, and Q. Zhang, Delay analysis of DRP in MBOA UWB MAC, in *IEEE International Conference on Communications, ICC'06*, vol. 1, pp. 229–233. IEEE, June 2006.

[9] R. Zhang and L. Cai, Optimizing throughput of UWB networks with AMC, DRP, and DLY-ACK, in *Proceedings of IEEE Global Telecommunications Conference*, 30 November–4 December, pp. 1–5, IEEE, 2008.

[10] K.-C. Go, J.-H. Kim, S.-H. Oh, K.-D. Moon, and K.-I. Lee, Resource allocation algorithm considering a priority of service classes for WiMedia UWB system, in *Proceedings of the 3rd International Conference on Ubiquitous Information Management and Communication*, ser. ICUIMC '09, pp. 298–301. ACM, New York, 2009. [Online] Available: http://doi.acm.org/10.1145/1516241.1516293.

[11] M. Daneshi, J. Pan, and S. Ganti, Towards an efficient reservation algorithm for distributed reservation protocols, in *Proceedings of IEEE INFOCOM*, March 2010, pp. 1–9. IEEE, 2010.

[12] K.-H. Liu, X. Shen, R. Zhang, and L. Cai, Performance analysis of distributed reservation protocol for UWB-based WPAN, *IEEE Transactions on Vehicular Technology*, vol. 58, no. 2, pp. 902–913, February 2009.

[13] K. Bartke-Minack and M. Guirao, On perceived throughput and delay fairness of a distributed reservation protocol, in *Proceedings of IEEE Wireless Communications and Networking Conference (WCNC)*, March 2011, pp. 31–36. IEEE, 2011.

[14] K.-H. Liu, X. Shen, R. Zhang, and L. Cai, Delay analysis of distributed reservation protocol with UWB shadowing channel for WPAN, in *Proceedings of IEEE International Conference on Communications (ICC'08)*, May 2008, pp. 2769–2774. IEEE, 2008.

[15] K.-H. Liu, X. Ling, X. Shen, and J. Mark, Performance analysis of prioritized MAC in UWB WPAN with bursty multimedia traffic, *IEEE Transactions on Vehicular Technology*, vol. 57, no. 4, pp. 2462–2473, July 2008.

[16] C. Hu, H. Kim, J. C. Hou, D. Chi, and S. S. Nandagopalan, Provisioning quality controlled medium access in ultrawideband-operated WPANs, in *Proceedings of 25th IEEE International Conference on Computer Communications (INFOCOM 2006)*, April 2006, pp. 1–11, IEEE, 2006.

[17] J. Robinson and T. Randhawa, Saturation throughput analysis of IEEE 802.11e enhanced distributed coordination function, *IEEE Journal on Selected Areas in Communications*, vol. 22, no. 5, pp. 917–928, June 2004.

[18] Z. Ning Kong, D. Tsang, B. Bensaou, and D. Gao, Performance analysis of IEEE 802.11e contention-based channel access, *IEEE Journal on Selected Areas in Communications*, vol. 22, no. 10, pp. 2095–2106, December 2004.

[19] Y. Xiao, Performance analysis of priority schemes for IEEE 802.11 and IEEE 802.11e wireless LANs, *IEEE Transactions on Wireless Communications*, vol. 4, no. 4, pp. 1506–1515, July 2005.

[20] J. Hui and M. Devetsikiotis, A unified model for the performance analysis of IEEE 802.11e EDCA, *IEEE Transactions on Communications*, vol. 53, no. 9, pp. 1498–1510, September 2005.

[21] X. Chen, H. Zhai, X. Tian, and Y. Fang, Supporting QoS in IEEE 802.11e wireless LANs, *IEEE Transactions on Wireless Communications*, vol. 5, no. 8, pp. 2217–2227, August 2006.

[22] Y. Lin and V. Wong, Saturation throughput of IEEE 802.11e EDCA based on mean value analysis, in *Proceedings of IEEE Wireless Communications and Networking Conference (WCNC 2006)*, April 2006, vol. 1, pp. 475–480. IEEE, 2006.

[23] X. Ling, K.-H. Liu, Y. Cheng, X. Shen, and J. Mark, A novel performance model for distributed prioritized MAC protocols, in *Proceedings of IEEE Global Telecommunications Conference (GLOBECOM'07)*, November 2007, pp. 4692–4696. IEEE, 2007.

[24] Omnet++, http://www.omnetpp.org/.

5

Resource Allocation and Energy Calculation in WPANs Based on WiMedia MAC

Muhammad Alam[1,2], Shahid Mumtaz[1], Christos Verikoukis[3] and
Jonathan Rodriguez[1]

[1]*Instituto de Telecomunições, Aveiro, Portugal*
[2]*Universidade de Aveiro, Aveiro, Portugal*
[3]*Telecommunications Technological Centre of Catalonia (CTTC), Barcelona, Spain*
e-mail: alam@av.it.pt

Abstract

The Distributed Reservation Protocol (DRP) guarantees channel allocation
for isochronous traffic, but an excessive use of DRP can lead to unfair resource distribution resulting in the incoming nodes to the system going into
backoff mode for long periods. On the other hand, PCA provides access to
the channel on a first come first served basis and can also lead to a similar
situation. Consequently, the devices in backoff mode use excessive channel scanning which leads to much reduced battery lifetime. This chapter
presents a solution for the resource allocation problem in WPANs based on
the WiMedia MAC. WiMedia provides a decentralized MAC protocol for
resource allocation, but still there are some issues related to the QoS provisioning which are highlighted in this chapter. Our study suggests that UWB
offers high data rates coupled with an energy consumption per bit which is
significantly lower compared to other short range technologies.

Keywords: Distributed Reservation Protocol, resource allocation, decentralized MAC, beacons.

Shahid Mumtaz and Jonathan Rodriguez (Eds.), Green Communication in 4G
Wireless Systems, 69–88.

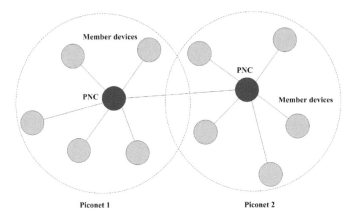

Figure 5.1 Piconet and inter-Piconet of 802.15.3a.

5.1 Introduction

MAC schemes for WPANs fall into two main categories: centralized and distributed. In the centralized approach, a central entity controls the activities of the network, such as resource allocation. An example of the centralized approach is the IEEE 802.15.3 protocol [1]. The IEEE 802.15.3 MAC forms a network called Piconet which consists of a central coordinator called a Piconet coordinator (PNC) as shown in Figure 5.1. The resource scheduling in the centralized approach is done by the PNC. The PNC collects the information about the member devices of the piconet and manages the demands for reservation; this information is then propagated to the rest of the members. IEEE 802.15.3 MAC offers a number of advantages such as, for e.g., high throughput in smaller networks, while also guaranteeing quality of service (QoS). One of the main advantages of using a centralized approach is to avoid collisions because the PNC controls access to the channel. The centralized MAC protocols are adopted for high bandwidth demanding applications such as video sharing and interactive gaming and are best suitable for small WPANs which are usually limited to 10 member nodes [2]. In the piconet each device accesses the channel in a time-division multiple access (TDMA) fashion while collocated piconets have to operate on different code channels to avoid interference, thus limiting the scalability of the centralized approach.

Undoubtedly a centralized approach offers a few advantages but it presents several problems, and the mobility challenge is one of them. For instance, if due to mobility the PNC goes out of reach of the member devices, the member devices will start excessive scanning to reach the PNC and consume a considerable amount of energy in doing so. Furthermore, if the PNC fails the whole network will go down and the re-election process will be initiated again to select a new PNC which will consume useful time and energy. This will also seriously affect the quality of service (QoS) especially for isochronous streams. Moreover, in the case of inter-piconet connectivity there are overlapping regions which degrades the efficiency of the IEEE 802.15.3. For instance, if two devices are in the overlapping regions and happen to use the same time slots, their transmission will collide and result in a degraded performance [3]. Also, the current IEEE 802.15.3 has difficulties with scalability to extend WPAN coverage.

On the other hand, in distributed MAC protocols there is no centralized controller and the resources are shared in a distributed way; WiMedia being a prime example that supports high data rate application [4]. This distributed approach eliminates the central point of failure and centralized resource management; devices now take their own decisions as shown in Figure 5.2. Moreover, the issue of scalability is resolved because the nodes can communicate with each other directly and thus can explicitly scale the network to a larger size, but will result other inherent problems such as synchronization and QoS provisioning. Problems like hidden terminals are resolved by using request-to-send and clear-to-end (RTS/CTS) mechanisms while the collision problems are minimized by using back-off and/or persistence mechanisms [5,6]. The WiMedia MAC manages the resources among the communicating devices via a reservation protocol called DRP and also using control and prob frames which are explained in next section. The reservation protocols and the control frames also solve the packet collision problem. The devices use a beaconing process to broadcast its presence and capabilities to the nearby devices in a periodic interval called a superframe. The beacons contain a number of information elements (IEs) which convey control and management information. A detail of UWB protocols and its comparison with short range technologies can be found in [7].

5.2 Beacon Frames

WiMedia is fully decentralized communication architecture and the beaconing process is used by the devices to uphold the organization of the network. In the resource allocation process the information regarding resources is communicated in the IEs of beacon. The devices first chose the channel and then each device shows its presence and communication capabilities to other neighboring devices with the help of Beacon frames. The Superframe has

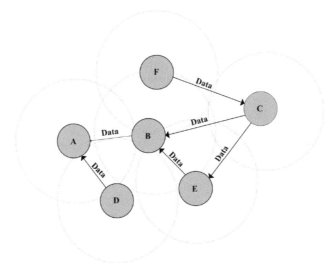

Figure 5.2 Piconet and inter-Piconet of 802.15.3a.

three main subdivisions that constitute: beacon period start time, beacon slots, and the media access slots as shown in Figure 5.3. The Beacon frames are sent in beacon slots within the Superframe by each device. In each Superframe the first two beacon slots are dedicated to a beacon signaling protocol. The beacon signaling protocol slots are followed by other beacon slots whose number depends on the number of active devices in a beacon group. The radio of each device starts to listen to the neighbor devices and if no device is transmitting its beacons then the new device begins transmitting beacons to setup network. The duration of the beaconing process is 10s and if no neighbor devices are found then the process must be stopped thus avoiding unnecessary signaling to trim down excessive use of energy and also to reduce noise. The length of each beacon slot is 85us and is fixed. The beacons are packed with information (e.g. which devices are nearby, what are their communication capabilities, what and how many MASs are reserved by these devices etc.) and by listening to these beacons the device will have a snapshot of the network. When a device wants to transmit a beacon it selects the available beacon slot not used by other devices. A brief description about the IEs in beacons is given in the next section.

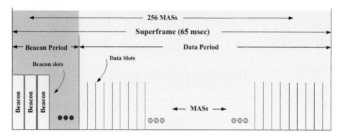

Figure 5.3 Superframe with beacon period.

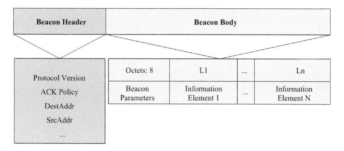

Figure 5.4 Beacon frame.

5.2.1 Information Elements (IEs)

The beacon frames are packed with a number of information items and each piece is called an Information Element (IE). The IEs that appear in the beacons, contain certain command and control frames that have the following general format, as defined in the ECMA standard and shown in Table 5.1.

Each IE carries a unique ID, and a length field that shows the length in octets followed by an IE-specific information field. As mentioned earlier, the beacons contains a number of information elements, some of the IEs and their IDs are shown in Table 5.2.

The "BPOIE" shows the information on the neighbours' BP occupancy in the previous Superframe. The "PCA Availability IE" shows a PCA availability bitmap and indicates the MASs that a device is available to receive PCA frames and transmit the required response. The length of PCA Availability Bitmap field is 256 bits. The "DRP Availability IE" gives a view to the device of the current utilization of MASs. The DRP Bitmap represents the availability of MASs and is up to 256 bit long and each MAS is represented by one bit in the Superframe. In order to save energy or in case the device does not want to participate in communication process, it goes into hibernation mode. This

Table 5.1 General format of IE.

Octet:1	1	1
BElement ID	Length	IE-Specific Field

Table 5.2 Information elements.

Element ID	Information element
1	Beacon Period Occupancy IE (BPOIE)
2	PCA Availability IE
8	DRP Availability IE
9	Distributed Reservation Protocol (DRP) IE
10	Hibernation Mode IE
11	BP Switch IE
12	MAC Capabilities IE
13	PHY Capabilities IE
14	Probe IE
18	Channel Change IE
21	Relinquish Request IE
3–7, 25–249, 253–254	Reserved
255	Application-Specific IE (ASIE)

news is shared with the neighboring devices by using the "Hibernation Mode IE". This IE specifies the number of superframes for which the device is going to be in sleep mode and also shows the wake-up time. The "BP Switch IE" indicates the device will change its BPST at a specified future time. The MAC capabilities of the devices are shown by the "MAC Capabilities IE", e.g. capabilities of PCA, Hard DRP, Soft DRP and Block ACK frames receiving and transmitting. The "PHY Capabilities IE" shows which PHY capabilities a device supports. The "Prob IE" indicates a device is requesting one or more IEs from another device or/and responding with requested IEs. When a device changes its channel, it is indicated by the Channel Change IE. "Relinquish Request IE" is used by a device to request other devices to release one or more MASs from one or more existing reservations. The frame contains the target device and the requested MASs with reason codes (e.g. if a target device holds more MASs than the permitted policy then this is represented as over-allocation).

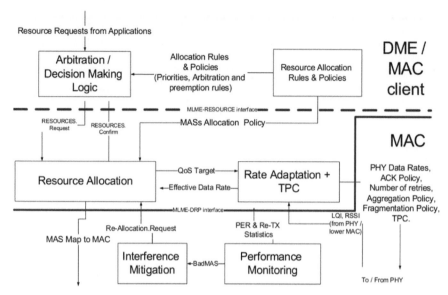

Figure 5.5 Resource allocation and rate adaptation reference model [8].

5.3 Distributed Resource Reservation in WiMedia MAC

For the resource allocation in WiMedia based WPAN, the WiMedia standard provided a reference model for the resource allocation and the rate adaptation architecture in [8] as shown in Figure 5.5. The application layer places a request for bandwidths based on the QoS of the application and bandwidth requirements, which are received at the MAC layer and shown as resource requests from application in the model. The MAC layer takes the decision regarding the requested bandwidths based on certain rules and policies, and the availability of the resources. The model also considers the rate adaptation rules, the interference mitigation and performance monitoring. Each device has its own bandwidth requirements based on the running application. As we mentioned in the previous section, the resource reservation is done via DRP to facilitate the devices the correct QoS.

The process of distributed resource reservation is accomplished via the DRP IEs in the beacon frames. The previous section explains WiMedia's beacons and IEs, here we present how the resource reservation process is done and which are the rules employed by DRP to reserve the MASs in superframe. The DRP IE is used to negotiate a reservation or part of a reservation for specified MASs and to announce the reserved MASs to the neighbor devices.

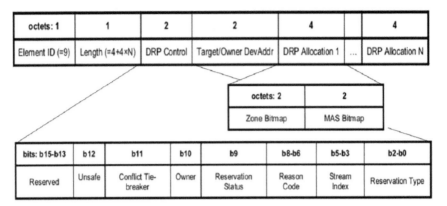

octets: 1	1	2	2	4		4
Element ID (=9)	Length (=4+4×N)	DRP Control	Target/Owner DevAddr	DRP Allocation 1	...	DRP Allocation N

octets: 2	2
Zone Bitmap	MAS Bitmap

bits: b15-b13	b12	b11	b10	b9	b8-b6	b5-b3	b2-b0
Reserved	Unsafe	Conflict Tie-breaker	Owner	Reservation Status	Reason Code	Stream Index	Reservation Type

Figure 5.6 DRP IE format with expanded view of DRP control and allocation fields.

The DRP IE is packed with a number of information items regarding the re-servation of resources as shown in Figure 5.6. The DRP control field contains a number of information it means related to a reservation request. Reservation types field specify the type of reservation e.g. Alien BP, Hard, Soft, Private and PCA. The type of data stream is represented by the Stream Index field. To indicate whether a DRP reservation request was successful, the response is shown via Reason Code field. If the DRP request is "successful" then the value of the reason code is set to zero. The different values for the Reason Code are set in the device if some of the MASs claimed in the reservation have been removed, or if DRP IEs have been combined, split or both. The Reason Code also shows whether a DRP request is pending, conflict, denied, modified or cancelled as shown in Table 5.3.

The Reservation Status represents the status of the DRP negotiation pro-cess and is set to ZERO in a DRP IE for a reservation that is under negotiation or in conflict, and is set to ONE by a device granting or maintaining a reser-vation. The field owner indicates the source or target of the DRP IE; it is set to ONE if the device transmitting the DRP IE or to ZERO if the device transmitting the DRP IE is a reservation target and reserved in case the Re-servation Type is Alien BP. The Conflict Tie-breaker bit is a random value set either Zero or One. All those DRP IEs that represent the same reservation, the Conflict Tie-breaker bit is set to the same value. If a reservation is considered in excess of the reservation limits, it is represented by the Unsafe bit and the value is set to ONE.

The Target/Owner DevAddr field represents the owner or target device. Its value is set to the target address by the owner if the owner is transmitting the

Table 5.3 Reason Code field encoding.

Value	Code
0	Accepted
1	Conflict
2	Pending
3	Denied
4	Modified
5	Cancelled
6–7	Reserved

DRP IE. The target address can be unicast or multicast. If the target is sending the DRP IE, it sets the devices address to owner device address. The field is reserved if the Reservation Type is Alien BP or PCA. The DRP IE contains DRP allocation fields that can be one or more. The DRP allocation is encoded using a zone structure under certain rules. In the zone structure the whole superframe is divided into 16 zones numbered from 0 to 15 starting from the BPST. Each zone contains 16 consecutive MASs, which are numbered from 0 to 15 within the zone. The zones that contain reserved MASs are represented by the Zone Bitmap. If a bit in the field is set to ONE, the corresponding MAS within each zone identified by the Zone Bitmap are included in the reservation, where bit zero corresponds to MAS zero within the zone.

5.4 Resource Distribution Problem

UWB offer a number of unique characteristics, but still there are a number of open challenges that need to be improved and enhanced at UWB MAC layer: distributed multiple access, distributed resource allocation, quality of service (QoS) provisioning, and overhead reduction [9, 10]. The components of QoS system are QoS mapping, admission control and resource allocation based on QoS requirement of each device [11]. Based on the traffic specification the WiMedia MAC cannot calculate and allocate air service rates to all the considered traffic streams, and the problem becomes more challenging with mobility and for two hop devices [11].

In [11], the authors highlighted the unfair resource allocation problem of WiMedia MAC. The example provided is shown in Figure 5.7. This figure shows five devices in a beacon group of DEV1. The traffic stream is represented by the arrow and the circle shows the transmission range of each device. The total bandwidth is 360 Mbps which is offered by 210 MASs. There are seven transmission streams (TS) created in a sequential order as shown in

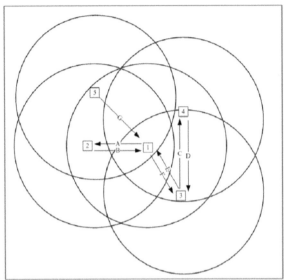

Traffic stream Index	Minimum service rate	Peak service rate	Service rate	CAMs (Current available MASs)
A(DEV1 DEV2)	30 Mbps	50 Mbps	50 Mbps	310 Mbps (After TS A is joined)
B(DEV1 DEV1)	50 Mbps	70 Mbps	70 Mbps	240 Mbps (After TS B is joined)
C(DEV3 DEV4)	60 Mbps	70 Mbps	70 Mbps	170 Mbps (After TS C is joined)
D(DEV4 DEV3)	20 Mbps	50 Mbps	50 Mbps	120 Mbps (After TS D is joined)
E(DEV3 DEV1)	35 Mbps	60 Mbps	60 Mbps	60 Mbps (After TS E is joined)
F(DEV1 DEV3)	50 Mbpsz	60 Mbps	60 Mbps	0 Mbps (After TS F is joined)
G(DEV5 DEV1)	30 Mbps	50 Mbps	Blocked	

Figure 5.7 Blocked service request of a new TS G (Total BW=360 Mbps for all 210 MASs) [11].

Figure 5.7 and after each TS the number of remaining MASs for other devices in the superframe reduces, and finally there are no MASs available for TS from DEV5 to DEV1. As a result, the TS from DEV5 to DEV1 are blocked.

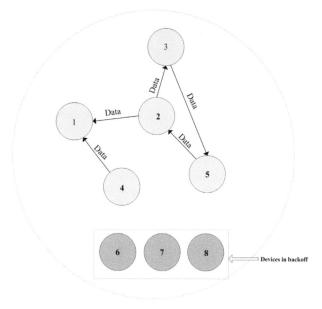

Figure 5.8 Example of resource allocation problem.

Let us consider another example; we have a connected network of 5 nodes in the beacon period as shown in Figure 5.8. The superframe is occupied by these five devices where DEV1-DEV3 uses the hard DRP and the DEV4 and DEV5 are using PCA. The final superframe is shown in Figure 5.9: there are a number of devices that want to join the network but due to the extended use of hard DRP, the available resources are not enough to accommodate the incoming devices into the system, thusrestricting the size of network. The incoming devices go to backoff and wait for the PCA using devices to release the MASs or DRP using devices to finish their complete data transfer. The same problem can arise in the previous example when some devices are reserving MASs by Hard DRP and the other has to wait in backoff mode.

The superframe final snapshot is shown in Figure 5.9. Due to Hard DRP reservation by DEV1-DEV3, the DEVs 6-8 are waiting for MASs to be released by DEV4 or DEV5.

Another situation which can degrade the throughput of MAC is presented in [12]. In the absence of a centralized controller, the devices access the channel and the reservation of the slots could be arbitrary. This can lead to a situation where the duration between the transmitting slots, which is called the vacation time, is not uniformly distributed in the scheduling cycles [12].

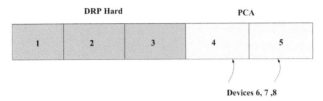

Figure 5.9 Example of resource allocation problem.

In [13], the impact of first come first serve based scheduling on fairness of throughput has been presented. As the nodes join the system on first come first serve basis they can occupy the channel for a longer duration and can significantly affect the throughput of system.

Another problem that can affect the throughput and increase the blocking probability is highlighted in [14]. The allocation of MAS should follow the order of isozones as much as possible. Each device in a single reservation can have eight MAS slots in a single reservation block. So if a flow cannot be allocated in one isozone the request is dropped to the next isozone of higher isozone index.

5.5 Solutions Provided to Resource Distribution Problem

A number of analytical and mathematical models have been presented in literature to provide a solution to the distributed resource allocation problem of WiMedia MAC protocol. In [11]the authors use traffic specification (TSPEC) presented in [15–17], for parameterized QoS and the fluid twin token bucket model presented in [8] to provide a simple solution to the resource distribution problem presented in the previous Standard terminologies are used to elaborate on the characteristics of a network traffic source, these are namely three parameters, mean rate r, peak rate p, and maximum burst size b, as shown in Figure 5.10.

The arrival curve of the cumulative maximum number of bits a source may inject in a time interval t represented in time t is shown in Figure 5.10. The following notations are used in [11] for the proposed mechanisms to solve the fair distribution problem:

K: Total number of TS_s that request for guaranteed QoS in the beacon group of a DEV;

MR_j: lower bounds of a service rate to guarantee QoS of TS_j;

PR_j: upper bounds of a service rate to guarantee QoS of TS_j;

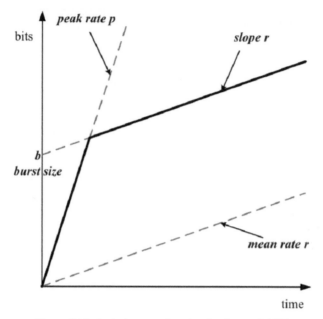

Figure 5.10 Arrival curve of a token bucket model [8].

$SR_{j,n}$: service rate allocated to TS_j at the nth superframe;
RR_j: number of MASs that TS_j relinquishes to accommodate more TS_s in a beacon group;
BW: denotes the total number of MASs forming the data period in a superframe.

The new introduced satisfaction ratio of QoS (SoQ_j) for the TS_j at the nth superframe is shown as follows:

$$SoQ_{j,n} = \frac{SR_{j,n} - MR_j}{PR_j - MR_j}$$

The SoQ of all the devices in a beacon period for all K and TS_s is given by the following equation and $SoQ_{F,n}$ has a value between 0 and 1.

$$\sum_{i=1}^{K}(SoQ_{j,n}(PR_j - MR_j) + MR_j) = \sum_{i=1}^{K}SR_{j,n} = BW$$

If the computed value of $SoQf_{F,n}$ is 1 then each TS_j in a beacon group is provided with PR_j which means that each transmission stream has peak rates

TS index	MR	PR	SoQ$_F$

Figure 5.11 QoS IE format.

and occupies maximum number of MASs. On the other hand if the computed value of $SoQ_{F,n}$ is 0 then each TS_j in a beacon group is provided with MR_j which means that each transmission stream has a minimum rate. A new IE is introduced based on the QoS parameters to exchange information about TS_s as shown in Figure 5.11.

The information included in QoS IE includes the TS index of TS followed by the minimum and maximum transmission rates and the SoQF value. Whenever the number of transmission stream varies in a beacon group, the changes are represented in QoS IEs and each device receives it in the nth superframe and then it calculates a new $SoQ_{F,n+1}$ for QoS provisioning at the next $(n+1)$th superframe. Each device has to decide whether to accommodate a new TS or not and how to allocate service rates to the existing TS_s and to the new TS. This decision is taken based on the calculated $SoQ_{F,n+1}$. If the new TS in a DEV requests a guaranteed QoS, then that device is allowed to put the QoS IE of the TS into its beacon. The same beacon group recognizes the request of the new TS, and they individually calculate $SoQ_{F,n+1}$. The proposed mechanism is evaluated based on two use-cases: in the first case TS with same token bucket TSPEC characteristics is evaluated, and in the second case TSs with different token bucket TSPEC characteristics are considered. A comprehensive view of fair distribution of resources is available in [11]. Figure 5.12 shows the actual throughput of each TS in its reserved MAS blocks for fair QoS provisioning according to D_SoQ (proposed mechanism) under varying traffic conditions.

A cross zone allocation and on-demand compaction solution has been proposed in [19]. The proposed algorithm can make reservations in multiple isozones. Contrary to the situation where a flow cannot be allocated in one isozone the proposed algorithm splits the request into sub-requests for the isozones of lower indexes including the current iso-zone. This cross-isozone allocation achieves higher system utilization as shown in Figure 5.13. The second algorithm that is presented in [19] is on-demand compaction similar to the technique used in memory management of operating systems. In this process all the allocated MASs blocks in each column of the isozone are pushed to one end of the column. A large free segment is achieved by com-

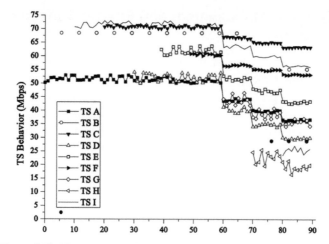

Figure 5.12 Measurement of TS_s throughput behavior in D_SoQ [18].

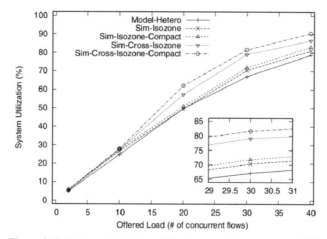

Figure 5.13 System utilization of the isozone-fit improvement [19].

bining all the free segments which extends to the other end of the column. This helps to utilize the superframe efficiently.

5.6 UWB Power Management and Energy Consumption Calculation

Energy consumption is not at the forefront of network design as most of these devices are battery powered, and applications become increasingly power hungry. To prolong the operation time of battery powered devices, the WiMedia standard manages the power through modes and states of the devices. An effective method to extend battery life is to enable devices to turn off completely or reduce power for long periods of time, where a long period is relative to the superframe duration [18]. The WiMedia standard also supports devices that sleep for portions of each superframe for power saving. A device may be in Active mode or Hibernation mode in each superframe. In Active mode, the device will send and receive beacon(s) in the current superframe, while in Hibernation mode the device will not send a beacon or other frames in the current superframe. When a device wants to enter the hibernation mode, it would have been advertised in previous superframe(s) that it is entering hibernation mode. Active mode devices may be in one of two power states within a superframe: Awake or Sleep. In the awake state, the device is able to transmit and receive while in sleep state the device does not transmit or receive. The power consumption of UWB is high, compared to the other short range technologies, e.g. Bluetooth and ZigBee as shown in Figure 5.14.

A study about the energy consumption of Bluetooth, UWB, ZigBee, and Wi-Fi is presented in [17] and WiFi, UWB and WiMax in [14]. The low power consumption of Bluetooth and ZigBee makes it more suitable for low data rate application compared to UWB and Wi-Fi. Figure 5.14 shows that UWB consumes more power relative to other technologies.

The energy consumption is usually calculated in joule per bit, as shown in the following formula:

Energy consumption per bit = Power consumption/Data rate

UWB supports large data rates up to 480 Mbps over a distance of 10 meters. Moreover, when we consider WiMedia, we have to take into account the context of the application, and in particular the inefficiency of the higher layers of the stack. Table 5.4 shows the throughput and the energy consumption for Wireless USB, with a simple but inefficient MAC layer, and with a customized MAC layer. We consider the average transmit power of UWB to be 555 mW [20]. Since UWB supports data rates up to 480 Mbps, we consider different data rates and calculate the energy consumed per bit.

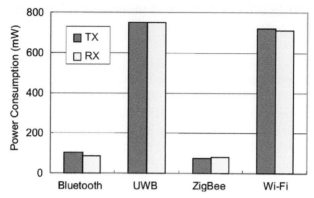

Figure 5.14 Power consumption comparison of different technologies [17].

Table 5.4 Energy Consumption for UWB [14].

Bandwidth (Mbps)	Throughput Wireless USB (Mbps)	Energy per Bit Wireless USB (nJ)	Throughput custom MAC (Mbps)	Energy per Bit custom MAC (nJ)
54	14.6	38.0	27	20.6
80	21.6	25.7	40	13.9
110	29.7	18.7	55	10.1
140	37.8	14.7	70	7.9
200	54.1	10.3	100	5.5
260	70.3	7.9	130	4.3
300	81.1	6.8	150	3.7
320	86.5	6.4	160	3.5
360	97.3	5.7	180	3.1
400	108	5.1	200	2.8
480	130	4.3	240	2.3

5.7 Conclusions and Future Challenges

This chapter has provided a survey on on WiMedia MAC resource allocation protocols. Although the WiMedia MAC works in a distributed manner and has solved many of the traditional issues related to using IEEE 802.15.3, there are still challenges with resource distribution which are highlighted in this chapter. Besides WiMedias DRP efficiency, distributed resource allocation is still an open challenge that needs to be addressed. The reservation process should be handled with proper balanced mechanism of the protocols (DRP-hard, DRP-Soft and PCA) proposed by the WiMedia alliance. The policies defined for the resource allocation affects the quality of service (QoS) provisioning of UWB based PANs. One of the main challenges that need to be addressed in future is the compaction of superframe. The devices reserve the MASs in the superframe in a row and column fashion, leaving a number of unused MAS and thus affect the QoS of the existing and incoming nodes to the system. This situation can also affect the QoE for incoming devices to the system by dropping packets or going to the backoff mode. Another challenge is to calculate and allocate fair service rate to all the considered traffic streams based on the specification of each traffic class. This fair service allocation problem becomes more challenging with mobility and multi-hop scenario. The ECMA standard defined a Physical and MAC layer protocol which offers a number of policies and control mechanisms to ensure quality of service (QoS) provisioning. A detail comparison of centralized and distributed resource protocols has been presented followed by details of WiMedia's MAC protocol and resource distribution protocol. The most typically encountered unfair resource distribution problems have been addressed here e.g. excessive use of Hard DRP, first come first served based approach to the resources etc. have been elaborated. Moreover, the existing solutions to the unfair distribution of resources based on existing state-of-the-art have been given. Our survey suggests that UWB offers more bandwidth and is suitable for high data rate applications and thus consumes less energy per bit compared to other short range technologies. Therefore, it can be considered a strong candidate for interactive gaming and other video and audio application. In this chapter we provided a short comparison of energy consumption of UWB with other short range technologies.

Acknowledgements

The research leading to these results has received funding from the European Community's Seventh Framework Programme (FP7/2007-2013) under grant agreement No. 248577 (C2POWER) and 23205 – GREEN-T, co-financed by the European Funds for Regional Development (FEDER) by COMPETE – Programa Operacional do Centro (PO Centro) of QREN. Muhammad Alam, PhD student with the MAP-I, would like to acknowledge the support of the grant of the Fundacão para a Ciência e a Tecnologia (FCT, Portugal), with reference number SFRH/BD/67027/2009.

References

[1] WiMedia MAC Release Spec. 1.01, Distributed medium access control (MAC) for wireless networks, December 15, 2006, http://www.wimedia.org/en/index/asp.

[2] W. C. Chung, N. August, and D. S. Ha, Signaling and multiple access techniques for ultra wideband 4G wireless communication systems, *IEEE Journal on Wireless Communications*, vol. 12, no. 2, pp. 46–55, April 2005.

[3] J.-W. Kim, K. Hur, J.-O. Kim, D.-S. Eom, and Y. Lee, A disturbed resource reservation structure for mobility and QoS support in WiMedia networks, *IEEE Transactions on Consumer Electronics*, vol. 56, no. 2, pp. 547–553, May 2010.

[4] C.-T. Chou, J. del Prado Pavon, and N. Sai Shankar, Mobility support enhancements for the WiMedia UWB MAC protocol, in *Proceedings of 2nd International Conference on Broadband Networks (BroadNets 2005)*, Vol. 2, pp. 136–142, October 2005.

[5] L. Kleinrock and F. Tobagi, Packet switching in radio channels: Part I – Carrier sense multiple-access modes and their throughput-delay characteristics, *IEEE Transactions on Communications*, vol. 23, no. 12, pp. 1400–1416, December 1975.

[6] T. Nandagopal, T.-E. Kim, X. Gao, and V. Bharghavan, Achieving MAC layer fairness in wireless packet networks, in *Proceedings of the 6th Annual International Conference on Mobile Computing and Networking*, ser. MobiCom '00, pp. 87–98. ACM, New York, 2000. [Online] Available: `http://doi.acm.org/10.1145/345910.345925`.

[7] M. Zin and M. Hope, A review of UWB MAC protocols, in *Proceedings of 2010 Sixth Advanced International Conference on Telecommunications (AICT)*, pp. 526–534, May 2010.

[8] ECMA International, 3rd edition. `http://www.ecma-international.org/publications/standards/Ecma-368.htm`, December 2008.

[9] X. Shen, W. Zhuang, H. Jiang, and J. Cai, Medium access control in ultra-wideband wireless networks, *IEEE Transactions on Vehicular Technology*, vol. 54, no. 5, pp. 1663–1677, September 2005.

[10] M. Zin and M. Hope, A review of UWB MAC protocols, in *Proceedings of 2010 Sixth Advanced International Conference on Telecommunications (AICT)*, pp. 526–534, May 2010.

[11] S. Kim, K. Hur, J. Park, D. Eom, and K. Hwang, A fair distributed resource allocation method in UWB wireless PANs with WiMedia MAC, *J. Commun. Networks*, vol. 11, no. 4, September 2009.

[12] K.-H. Liu, X. Shen, R. Zhang, and L. Cai, Performance analysis of distributed reservation protocol for UWB-based WPAN, *IEEE Transactions on Vehicular Technology*, vol. 58, no. 2, pp. 902–913, February 2009.

[13] K. Bartke-Minack and M. Guirao, On perceived throughput and delay fairness of a distributed reservation protocol, in *Proceedings of IEEE Wireless Communications and Networking Conference (WCNC)*, March 2011, pp. 31–36. IEEE, 2011.

[14] M. Albano, M. Alam, A. Radwan, and J. Rodriguez, Context aware node discovery for facilitating short-range cooperation, paper presented at 26th Wireless World Research Forum (WWRF26), April 2011, Doha Qatar. Available at `http://www.di.unipi.it/\~michele/papers/Albano\_WWRF\_WG5\_final.pdf`.

[15] S. Shenker and J. Wroclawski, RFC 2215 general characterization parameters for integrated service network elements. [Online] Available: `http://www.faqs.org/rfcs/rfc2215.html`, September 1997.

[16] R. Zhang and L. Cai, Optimizing throughput of uwb networks with AMC, DRP, and DLY-ACK, in *Proceedings of IEE Global Telecommunications Conference (GLOBECOM 2008)*, 30 November–4 December 2008, pp. 1 –5. IEEE, 2008.

[17] J.-S. Lee, Y.-W. Su, and C.-C. Shen, A comparative study of wireless protocols: Bluetooth, UWB, Zigbee, and Wi-Fi, in *Proceedings of 33rd Annual Conference of the IEEE Industrial Electronics Society (IECON 2007)*, November 2007, pp. 46–51. IEEE, 2007.

[18] J. Pavon, S. S. N, V. Gaddam, K. Challapali, and C.-T. Chou, The MBOA-WiMedia specification for ultra wideband distributed networks, *IEEE Communications Magazine*, vol. 44, no. 6, pp. 128–134, June 2006.

[19] M. Daneshi, J. Pan, and S. Ganti, Towards an efficient reservation algorithm for distributed reservation protocols, in *Proceedings of IEEE INFOCOM 2010*, March 2010, pp. 1–9. IEEE, 2010.

[20] O. Grondalen, P. Gronsund, T. Breivik, and P. Engelstad, Fixed WiMax field trial measurements and analyses, in *Proceedings of 16th IST Mobile and Wireless Communications Summit*, pp. 1–5, July 2007.

6

Clustering Techniques for Energy Efficient Wireless Communication

Victor Sucasas[1,2], Hugo Marques[3], Jonathan Rodriguez[1] and
Rahim Tafazolli[2]

[1]*Instituto de Telecomunições, Aveiro, Portugal*
[2]*Universy of Surrey, Centre of Communication System Research, UK*
[3]*Instituto Politecnico de Castelo Branco, Portugal*
e-mail: vsucasas@av.it.pt

Abstract

The increasing set of features and hardware capabilities of new generation
mobile devices, equipped with wireless interfaces such as Bluetooth and
Wi-Fi, has opened new paths for energy efficient wireless communication.
Nevertheless, stability and scalability in ad hoc networks must be provided
before taking advantage of energy saving in short range communication, and
clustering algorithms are envisioned for that task. In this sense, among the
clustering algorithms in the state of the art, mobility-aware algorithms are
considered the best solution for the dynamic nature of ad hoc networks in
urban scenarios. Although there are many proposals for clustering algorithms
in mobile networks, an adaptation is needed to be adapted for urban characteristics, where the mobility pattern and the environmental conditions have a
big impact on the algorithm performance.

Keywords: Clustering, ad hoc networks, energy efficiency, green.

Shahid Mumtaz and Jonathan Rodriguez (Eds.), Green Communication in 4G
Wireless Systems, 89–115.

6.1 Introduction

The mobile ad hoc networks (MANETs) were originally designed for a reduced set of scenarios, where the infrastructure-based topology is assumed not to be deployable. Among this set [1] were military communications in the battlefield and emergency applications (natural disasters, rescue operations, etc.) nevertheless, the increasing capabilities of mobile handsets in terms of short range wireless communication, such as Wi-Fi and Bluetooth, has led to a vision of reliable and efficient MANETs powered by cooperative communications.

The field of Wireless Sensor Networks (WSNs) is also an example of how efficient communications can be accomplished in infrastructure-less scenarios. Sensor networks are formed by small devices with reduced processing capabilities and limited and not-renewable battery. These sensors are deployed to gather data that is later forwarded to a common base station to be centralized and processed. In such scenario, the energy cost of the communication is a critical issue regarding the network lifespan, and cooperative techniques are usually proposed to decrease the energy used to forward the data from the sensor to the base station. Sensors perform a multihop delivery of data to the centralized database by using short range communications. This ensures connection with sensors that could not contact the base station in only one hop and saves energy by using short range communication, instead of a longer range communication that some nodes would have to perform to reach the base station in one hop. Typically, neighbouring nodes cooperatively team up to send data to relaying nodes that forward such data to the base station. Thus, only a reduced number of sensors are chosen to transmit directly to the base station.

In the same way as WSNs, users of PDAs, tablets and smartphones can establish a short range collaboration to locally exchange data whilst one or a few of the devices act as relay agents towards the long range connection with the service providers, hence achieving a more energy efficient communication. As an example, in [2] a clustering structure is proposed to team up mobile devices using Bluetooth interfaces, see Figure 6.1. The proposal experimentally probes the efficiency of using Bluetooth among nodes organized in clusters that can dynamically adapt the cluster to the application requirements of every node. Mature work and detailed specifications in this field can be also found in C2POWER project [3], that uses cognitive radio and cooperative strategies to take advantage of energy efficient short range communications. Cooperative approaches based in MANETs can also be en-

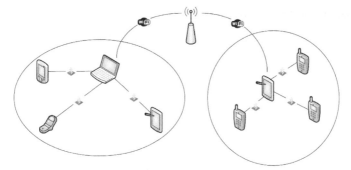

Figure 6.1 Multi wireless interface clustering architecture for energy efficient communications using Bluetooth for short range and low data rate, and Wi-Fi for long range.

visaged like a way of improving network features. Nodes into collaborative teams can commit their resources into a communal pool of resources that are fairly shared among the nodes in the team, allowing nodes to get capabilities they are not equipped with [4]. Besides, some features emerge from cooperation since the mutual collaboration of more than one terminal is needed for some applications (i.e. localization systems not based on GPS).

Apart from the mentioned applications other active fields of research (user or channel diversity, cooperative relaying, etc.) are based in cooperative communications, thus providing algorithms to sustain ad hoc networks communications is a must. Although cooperative techniques require more considerations than the topology distribution, in this chapter we intend to provide a detailed description of clustering techniques, and summarize the benefits and drawbacks of the most known clustering algorithms.

6.2 Benefits and Cost of Clustering

6.2.1 Benefits of Clustering Architectures

Due to the dynamic nature of mobile ad hoc networks, the research community has persistently developed new routing protocols. Some of these protocols were adaptations of well-known wired routing protocols modified to support the fast variations of this kind of networks; such modifications included on-demand route discovery or explicit control messages for topology information exchange.

Most of these proposals however, are not scalable over a flat topology [5]. Routing protocols are categorized by being reactive or proactive regarding the periodicity and the quantity of information gathered by nodes on the topology of the network. In proactive routing protocols (DSDV, OSLR), nodes exchange information periodically to have a global knowledge of the network regardless of which routes are really needed, thus new connections can be rapidly established using the information previously gathered. The main drawback of these protocols is the scalability, the information kept by each node increases exponentially with the increase of the number of nodes in the network. On the other hand, reactive routing protocols (AODVv2, LOADng, DSR, TORA) only keep information of the routes that are needed for communication in higher layers. Every time a new path towards a destination is needed, the route must be discovered by a route request and its reply from the destination. Although the size of the network does not affect the quantity of information stored in nodes, large networks introduce bigger delays at the time of route discovery.

To reduce these effects, hybrid approaches suggest using a proactive routing protocol for a reduced part of the network and cover the rest of the network with a reactive protocol; this enables nodes to have more detailed routing information on close neighbours that are just a few hops away. The number of hops determines the size of the area covered by every node and can be dynamically adapted.

Clustering architectures are proposed to overcome the scalability problem. This technique consists on choosing a smaller subset of nodes in the network through which the packets are routed. Limiting the number of possible routes in the network makes the routing information considerably smaller. The nodes elected to form this backbone are the leaders, commonly designated by Cluster Heads (hereby defined as *clusterheads*). The rest of nodes in the network, cluster-members, join one of these *clusterheads* to form virtual groups of nodes, called clusters. Normally inter-cluster communications are routed through this backbone, and the cluster-members do not need to keep information about the rest of the nodes outside their own cluster. Additionally, some cluster-members that are in transmission range with other clusters are used as gateways to connect the *clusterheads* and form the backbone. A typical wireless network using a clustering architecture is depicted in Figure 6.2.

The routing protocol deployed for inter-cluster communication is independent from the clustering topology, thus proactive or reactive routing protocols are allowed. Reactive protocols suit better if the topology changes frequently and proactive protocols are elected for more stable networks where the nodes are static.

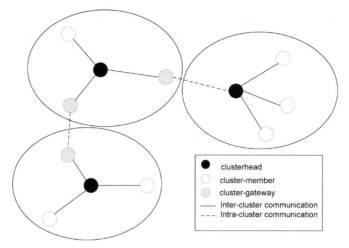

Figure 6.2 Common clustering architecture. The circle represents a cluster, not necessarily the transmission range of a clusterhead.

Furthermore, the division of the topology in clusters allows for a more efficient channel or code allocation. Clusters can reuse channels when they are not overlapped. Regardless the channel access method that the nodes use FDMA or CDMA, the number of channels or codes is limited. Dividing the topology in smaller groups aids the implementation of colouring algorithms for smart channel allocation to avoid interference. CDMA is commonly found in the literature for inter-cluster communication, where clusters are assigned different orthogonal codes. Although the spectral efficiency is the same in comparison with FDMA, CDMA allows asynchronous deployment. The clusters compute pseudo-random codes that are probabilistically orthogonal, thus we do not need a colouring algorithm assigning codes to clusters. The drawback is that the more neighbouring clusters a given cluster has the higher the probability of interference between them.

With respect to intra-cluster communication, *clusterheads* and cluster-members can be easily coordinated (they are connected in one or few hops), thus TDMA strategy can be deployed. The *clusterhead* is usually proposed to choose this strategy and assign timeslots to cluster-members for transmission, avoiding interference in intra-cluster communication. In [6] intra-cluster strategy, nodes have timeslots to transmit and there is always a free timeslot to listen for newcomers willing to join the cluster. All the communications in the cluster are done using the given channel or code assigned to the cluster, except for the explicit control communication used to form and maintain the topology. Control communication uses a common control channel for all clusters. In this approach, the nodes in the cluster take turns to advertise the cluster periodically. In other proposals nodes periodically broadcast a hello messages including information about themselves and the cluster they belong to; the latter is used for topology formation and maintenance. There is a trade-off about the periodicity and quantity of information to be sent through this common control channel; the more frequent the information is sent the more adaptable the algorithm is to changes but the less energy efficient in terms of transmitted overhead.

Other emerging capabilities in clustering architectures are related with task sharing. In a flat topology the nodes can cooperate to use each other's hardware or software capabilities but the coordination needed is not easy to accomplish. In a cluster environment the coordination is already provided, aiding the cooperation of nodes that can use other's interfaces or hardware equipment remotely. Multi-parallel processing is an example of how a high-demanding task can be shared by several terminals and perform faster. Other mechanisms like localization mechanisms [7, 8], where the position of some nodes can be calculated from the position of few nodes equipped with GPS-system, also need synchronization among several terminals. Clustering architectures allow coordination in intra-cluster communication, helping the deployment of this kind of mechanisms.

6.2.2 Costs and Design Considerations in Clustering Architectures

Although clustering architectures provide a good set of improvements, there are also drawbacks that must be considered when designing a clustering algorithm [9]. Such penalties can be listed as: (i) explicit control information exchange; (ii) ripple effect; (iii) stationary assumption; (iv) computation rounds and time for completeness; and (v) energy consumption.

These drawbacks must be carefully studied when designing a clustering algorithm. The environmental conditions regarding the network in which the clustering algorithm is to be applied has a high impact in the performance of the algorithm. As an example, a static network allows more complex algorithms for the optimization of the final topology regarding a predefined parameter (i.e. the length of the backbone or the size of the clusters). In such networks, the number of rounds or time that the algorithm consumes is not critical when in comparison with a mobile network, where the changes occur so frequent that spending time in optimization makes no sense. Adapting the algorithm to the conditions of the network and the requirements of the users is a necessity, and assessing the limitations of the network is also critical (some parameters, such as a high QoS demand, can be difficult to achieve in some MANETs).

The first drawback mentioned is the explicit control information exchanged by the nodes for cluster formation and maintenance. The nodes must exchange information with neighbours to advertise themselves, to let the neighbours be aware of their presence in the network and form part of the topology. These messages also include information about the neighbourhood to let other nodes know about the topology out of the scope of their transmission range. The status of the node in the topology and other specific information used by the clustering algorithm is also exchanged in these messages. Information about the neighbourhood is needed for inter-cluster communication. The *clusterhead* of a cluster obtains information about other clusters directly from the members of those clusters or from its own members that detect the presence of nodes belonging to other clusters. These members are to be elected as cluster-gateways for inter-cluster communication. Thus, a *clusterhead* should be aware of the neighbouring clusters. The information about the rest of the clusters in the topology (not connected to any gateway) is obtained using routing protocols through route discoveries or other routing techniques. Clustering algorithms are placed between the network and MAC layer, but are independent from the routing protocol deployed.

Information about the status of a node is known once the topology has been formed. A node must specify its role in the network as a cluster-member, *clusterhead* or gateway. Besides, clustering algorithms use node related information to elect *clusterheads* and form clusters. This information can be very basic (i.e the unique identifier of the node) or computed from node information (i.e mobility, degree, pattern consumption or even a combination of several parameters). This information is included in the control messages. As mentioned previously, nodes have a common control channel or code to send this information, and periodically send the so called "hello message" including this information. The strategy adopted to send this information regarding the periodicity and the quantity of information (i.e. it can be split in several kinds of messages) can vary depending on the clustering algorithm proposed.

The ripple effect happens when a simple change in the network causes significant changes throughout the network. This is common in algorithms that perform a strong optimization of the topology and that strictly follow some rules that cannot be broken after the cluster formation, in the maintenance phase. Such rules are commonly deployed in static networks and Dominated Set based algorithms is one of such situations that performs several rounds applying "marking" and "deleting" rules for obtaining the minimum backbone for a fully connected to have all network connected. Although these algorithms usually do not have maintenance phase, and re-clustering is performed periodically to adapt the topology to the changes, any given change in the position of a node can completely change the topology requiring an update for all routing information. This effect also happens in other algorithms that have a maintenance phase. It is usual in 1-hop clustering when a node moves out of the range of its current *clusterhead* and cannot reach other *clusterhead*; triggering an adaptation of the topology to connect this node, leading sometimes to the restructuration of the whole topology. Some algorithms avoid this phenomenon by transforming these nodes in *clusterheads* (with no members) or joining them to other clusters in a 2-hop connection [10].

As will be described in Section 6.3, the information used for cluster formation can include hardware or software capabilities, energy (i.e. battery level or pattern consumption), mobility of the node, position, distance w.r.t. neighbours, degree of the node (number of neighbours), amongst others. This information is periodically updated and sent, and is an important aspect to consider in a clustering algorithm. Except in few cases, where the information does not change (i.e. the identifier of the node), the validity of the information sent has an expiration time. The time that the algorithm takes to form the topology using this information must be in concordance with the time of validity of such information. As an example, a clustering algorithm based in the current position of the nodes, where nodes are moving 5 m/s and the algorithm takes 10 seconds to complete the topology is useless. Using information about the speed and direction of a node instead of its position could attenuate the previous problem but has the drawback that we could select *clusterheads* with low battery level. We must take into account that, apart from the expiration time, the validity of the information used can depend on other not related information (like mobility and battery level as in the previous example).

Other design considerations that need to be taken into account when designing an algorithm is the topology. The number of clusters in the network and the number of members per cluster directly affects the performance in inter and intra-cluster communications. The less the number of clusters the smaller the backbone formed between *clusterheads* and gateways and the less the hops for inter-cluster communication, which aids routing protocols to reduce control information and perform better (i.e. in terms of delay). On the other hand, if we reduce the number of *clusterheads* we implicitly increase the number of members per cluster, leading to overloaded *clusterheads* that drain their battery faster and where the members have less timeslots for intra-cluster communication.

Although the type of wireless technology in use and the kind of applications used by nodes inside the cluster can have different transmission requirements, there is always an optimum number of members per cluster that optimize this trade-off. The number of hops allowed in a cluster is also a design decision. The performance (in terms of throughput and delay) of an ad hoc network highly decreases with the number of hops [11], and 1-hop clustering is advised when we must provide the best QoS. However, the mobility of nodes with 1-hop clustering can easily lead to disconnection of nodes from the *clusterhead* and continuous re-affiliation. In some cases when the nodes move out from the transmission range of any *clusterhead* in the network some algorithms perform the clustering algorithm again. They re-select a new set of *clusterheads* to cover all nodes, provoking what was previously described as ripple effect. A less disturbing decision would be selecting these lonely nodes as *clusterheads* that would temporarily have no members.

In [12] the authors proposed a measure to determine the load balancing of a clustering architecture regarding the number of members per cluster:

$$LBF = \frac{n_c}{\sum_i (x_i - \mu)^2} \tag{6.1}$$

where n_c is the number of *clusterheads*, x_i the number of members of the cluster i, and $\mu = (N - N_c)/n_c$ the average of cluster-members per cluster. N is the total number of nodes in the system. This metric is higher when the clusters are balanced, having similar number of members. We should also take into account that the number of clusters cannot be arbitrarily set. If σ is the optimum number of members per cluster then the optimum number of clusters is $n_c = N/(\sigma + 1)$.

We must also take into account that the election of clusterheads based on the optimization of a given metric is a NP-hard problem [12] and can only be approximated by heuristics. The next section will describe several heuristics for clusterhead election.

6.3 State of the Art in Clustering Algorithms

In [13] the authors proposed a clustering algorithm based in the unique identifier of the nodes; selecting *clusterheads* based uniquely uniquely in the ID of the node. The lowest ID in the neighbourhood sets itself as *clusterhead* while the neighbouring nodes join them and do not participate in next rounds. The rules set by this algorithm are usually taken as a base by many other proposals for 1-hop clustering that use more complex metrics instead of the ID of the node. The main drawback of this algorithm is the unfairness of selecting always the same nodes as *clusterheads*, leading to a fast drainage of their batteries. The highest degree algorithm [14] does not fall in the same mistake since it forms the *clusterhead* set with the nodes with more number of neighbours. Thus, it minimizes the number of clusters. As stated before this decision has the side effect of maximizing the number of members, with the risk of overloading the *clusterheads*.

Unlike previous algorithms, Dominated Set based clustering techniques have a more complex set of rules to optimize the election of *clusterheads* in such a way that the size of final backbone is minimized. In a connected dominated set [15] *clusterheads* are directly connected forming a backbone that is reachable in 1-hop by all the rest of the nodes in the topology. This requires a double-phase process where suitable nodes are marked as candidate *clusterheads*. In the following rounds some of these nodes are deleted form the backbone and the algorithm is periodically performed to maintain a connected topology. A weakly dominated set [16] allows the *clusterheads* not to be directly connected and form interconnected pieces (groups of nodes). These algorithms are only suitable in static networks due to the re-clustering process that must be performed with every single change.

Low-maintenance clustering is based in minimizing the changes in the topology after cluster formation. First proposals maintained the topology by periodically re-clustering the network, introducing a considerable overhead. LCC (Least Cost Clustering) [17] is considered the first one suggesting the division of clustering techniques in two phases, formation and maintenance. Once the cluster is formed the changes are minimized, not allowing the nodes to change the cluster, and reducing significantly the cases where the re-clustering is performed. Although it is considered that the best solution is having the same rules for formation and maintenance phase, this does not mean that the algorithm should be performed from time to time. Minimization of overhead can be accomplished by applying the changes locally where the change is produced thus avoiding ripple effect. In this sense merging and splitting processes can be applied to join two *clusterheads* when they get in the transmission range of each other, or when the size is overpassing a predefined bound. Other heuristic that could be included in this category is PC (Passive Clustering) [18] where no explicit control information is exchanged. The information needed for cluster maintenance is added in data packets as overhead. While the communication is in course the nodes are learning about the topology. The problem is that when a node is not transmitting for some time it is deleted from the topology. This protocol could only be deployed if the nodes have a continuous data transmission.

For mobile ad hoc networks there are many proposals related with mobility-aware metrics, where the speed and direction of nodes is taken into account to elect the *clusterhead* set. In a highly mobile environment mobility is the most determinant parameter for the cluster stability. A more detailed overview on this kind of clustering techniques will be given in Section 6.4. However, energy-efficient techniques, load balancing and composit metrics also deserve the attention of this chapter and are described in this section.

Energy-efficient techniques are well suited for wireless sensor networks where the topology is frequently static. Sensors are usually deployed with a limited and rechargeable battery, and it is of most importance to minimize the energy consumption. Defining the lifespan of a network as the time while all nodes are still alive, these algorithms try to increase this metric by, for example, fairly sharing the role of *clusterhead*. Sensors are deployed to gather data that is later sent to a base station. Using short range communications they send this information to their *clusterhead* that is in charge of deliver it to the base station. Thus, receiving the information and sending it in long range communication is an energy consuming task that should be rotated. LEACH [19] uses random numbers to fairly rotate the *clusterhead* role. The HEED protocol [20] uses battery level as a primary parameter and the communication cost as a secondary parameter to compute the probability that the node is elected as *clusterhead*. The probabilistic fashion ensures some sort of randomness where the nodes with higher battery level and less communication cost (in terms of distance to the neighbours and number of neighbours) are more likely chosen as *clusterheads*.

Load-balancing clustering deals with the size of clusters as mentioned in Section 6.2. In [21, 22] bounds are set regarding the size of the cluster, merging and splitting procedures are suggested to maintain the size inside the allowed interval. This solution however, lacks the consideration of transmission range of nodes and possible inter-cluster interference. An example that depicts this problem is a overwhelmed cluster in a dense network where all nodes can hear each other. By splitting this cluster, it results in several overlapped clusters that must deploy different channels to avoid interference. Colouring algorithms for channel allocation in a dynamic network is not an easy task and asynchronous CDMA, where clusters can compute themselves pseudo-random orthogonal codes, do not ensure a complete avoidance of interference. Thus, it is advisable that in dense networks the size of the clusters is controlled considering (fine tuning) the transmission range of the nodes.

On another approach Chatterjee et al. [12] take into account several parameters. The authors proposed using a combination of mobility M_v, degree (number of neighbours) v, pattern consumption of the battery P_v, and distance to the neighbours D_v. These parameters are summed in a weighted manner:

$$W_v = c_1 \Delta_v + c_2 D_v + c_3 M_v + c_4 P_v \qquad (6.2)$$

Each node v computes its metric W_v. The ones with the lowest metric in the neighbourhood are elected as *clusterheads*.

The weight factors allow the prioritization of some parameters in the final metric, which is useful for adapting the algorithm to the requirements of the network. In the distributed version of the algorithm the final metric is the only information sent by the nodes in the control messages and the nodes with the best metric in the neighbourhood are elected as *clusterheads*.

A more optimal set of *clusterheads* can be obtained with optimization algorithms. In [23] the authors used a Tabu Search optimization to find the *clusterheads* by looking iteratively for the solutions that have the best outcome in a predefined fitness function. The election of the weights however, is not a straight-forward task, and some parameters can completely obfuscate the value of a given parameter. As an example, a static node with an optimum number of neighbours can get a good final metric even if its battery is almost depleted. This effect can occur if more weight is given to mobility and degree parameters and a lower one for consumption pattern. On other approach, Cheng et al. [24] used a multi-object optimization technique to select *clusterheads* using directly the mentioned parameters (mobility, degree, distance and consumption) instead of the metric described. It is based in an evolutionary algorithm that iteratively computes random possible *clusterhead* sets and picks the pareto-optimal solutions. After some predefined iterations the algorithm searches for better solutions using mutation and crossover techniques. Although these optimization algorithms lead to a near optimal election of the *clusterhead* set they require a global evaluation on the information of all nodes, thus centralization of the algorithm is a requirement. Furthermore, it also shares the obfuscation problem mentioned above.

6.4 Mobility-Aware Clustering Algorithms

Mobility-aware algorithms mainly use the mobility parameter as the most relevant information to elect the *clusterheads* in the topology. This information can be represented in many ways; the change of the position of a node in a given time, the current speed and direction of the node, the variations of the signal strength or the frequency (Doppler effect). Recently the most prolific field in this area is related with position and mobile information obtained by GPS systems, since it is becoming a common feature in smartphones, tablets, laptops and PDAs. However such solutions are not energy efficient and are also significantly limited in indoor environments.

Algorithms like MOBIC [25], based on signal strength or frequency variations in two consecutive control messages to assess stability of nodes, do not have the energy cost associated to GPS but multipath or fading effects can lead to misinterpret the mobility of a node. Other proposals explained along this section try to overcome these drawbacks by not using information related with signal received or GPS information.

6.4.1 GPS-Based Solutions

As stated, in [12] the authors used several parameters to assess the node's suitability of a node to become *clusterhead*. Concretely, the mobility parameter M_v is calculated by gathering the current location of a node at different times:

$$M_v = \frac{1}{T} \sum_{T}^{1} \sqrt{(X_t - X_{t-1})^2 + (Y_t - Y_{t-1})^2} \qquad (6.3)$$

The lower the mobility metric the more suitable the node is to become *clusterhead*. This simplistic metric can be deployed in a network with low mobility. However, for highly mobile networks this algorithm cannot perform well. Moreover, WCA algorithm is composed by several parameters and mobility is not commonly used as the most important. Load balancing (related with the number of neighbours parameter) has the most weigh in the metric calculation for this protocol.

Other algorithms use the group mobility feature present in human walk mobility, where nodes are supposed to share destination and speed with some neighbouring nodes, to form groups. In [26] the authors proposed the Distributed Group Mobility Adaptive Clustering algorithm that gathers the location of the node in discrete times. If the last sample differs in more than a predefined value from the previous one it computes the direction and the speed of the node using the new values. By comparing speeds with the neighbours, every node computes the "Spatial Dependency" that is the level of similarity of the speeds and directions of a pair of nodes. The Total Spatial Dependency (TSD) is the addition of the Spatial Dependency (SD) of each neighbour, thus the higher the TSD implies the node has bigger set of neighbours with similar mobility and it is considered a good candidate to be *clusterhead*.

In an SCP clustering scheme [27], a new metric based in the speed of the nodes is proposed. Every node obtains its speed defined in two coordinates. By gathering information from the neighbouring nodes they compute the average speed in the neighbourhood and assess their own stability by comparing their speed with the average:

Speed of the node:

$$v_i = v_{xi} + v_{yi} \tag{6.4}$$

Difference w.r.t. the average:

$$\Delta v_i = \sqrt{(v_{xi} - \bar{v}_x)^2 + (v_{yi} - \bar{v}_y)^2} \tag{6.5}$$

Stability factor:

$$S_i = \frac{1}{\Delta v_i + 1} \tag{6.6}$$

Willingness factor:

$$W_i = 2^{(S_i \log_2(M_i+1))} - 1 \tag{6.7}$$

The willingness factor, where M_i counts the number of neighbours with similar mobility w.r.t. the node i, is used as a metric to select the *cluster-heads*. The bigger the willingness factor the more adequate the node to get the *clusterhead* role. The algorithm introduces a threshold for the willingness factor in the maintenance phase to assess if a *clusterhead* must resign its role. This way a *clusterhead* leaving a group can resign before losing contact with the members. The value of this threshold is a trade-off that must be carefully assessed according network characteristics. A high value would lead to frequent *clusterhead* re-election and a low value could not adapt the topology to changes.

In MBC [28] the authors used the vector of velocity of each node to compute the Relative Mobility of each pair of nodes n and m:

$$M_{m,n,T} = \frac{1}{T} \sum_{i=1}^{T} |v(m, t) - v(n, t)| \tag{6.8}$$

The algorithm proposes a hierarchical construction of the topology where two metrics are proposed: "Cluster Mobility " (CM1) is used to assess the speed and direction average of the Cluster, and "Cluster Mobility 2" (CM2) to assess the uncertainty of the mobility of nodes inside the cluster; the size of the cluster is adapted according to these two metrics. The hierarchical construction indicates that not all the *clusterheads* in the topology are included in the backbone but only the top level *clusterheads* that are more stable will fill that role. The other *clusterheads* are hierarchically tied to the previous ones.

A centralized solution that can accomplish good accuracy with low rate of information gathering is proposed in [29]. Nodes update information in a server that is considered to be in the origin of coordinates. Each node (moving object "O") is described by $O(OID, x, v, t)$, representing respectively identity, position, velocity and time of the sample. Clusters are also described with a data set $CF = (N, \mathbf{CX}, \mathbf{CX}^2, \mathbf{CV}, \mathbf{CX}^2, \mathbf{CXV}, t)$.

This data set is composed by the number of nodes N, the addition of position of nodes and its squares, velocities and its squares and the addition of pairs of positions and velocities respectively. The addition of one node to the cluster does not imply a recalculation of the "CF" value of that cluster, only the addition of the values of the node "O" to the "CF". The decision of adding nodes to a cluster or forming new clusters is made using "CF" and "O" data sets, computing the distance between the nodes and the geographical center of the cluster. The main advantage is that with the velocity, position and time of the last actualization of nodes and clusters, it is possible to compute new values and update the data sets even if information is not received during some time. This allows low rate of sample gathering. The main drawback is the centralization of the algorithm and the overhead associated with the exchange of associated data.

The KCMBC algorithm [30] uses a different approach. Among other parameters, this algorithm uses the mobility of nodes (position and speed) to predict expiration time of links between nodes. Each node compares the speed with each neighbour and assesses the expiration time of the link with that neighbour. Once a node has predicted the expiration time of all its links, the average is calculated and compared with the average of its neighbours. This metric is used to elect the nodes with more stable links to be *clusterheads*. This algorithm also proposes a dynamically adaptive broadcast period to exchange information according to the velocity of the nodes.

If the network is somewhat stable (slow mobility) it is possible to use the position of the nodes instead of the speed and direction. In [31] the authors used GPS information to split the scenario in a grid. The *clusterhead* is the closest node to the center of the grid, for each of the different areas in the grid. When the *clusterhead* becomes unavailable the closest node to that *clusterhead* is chosen as the new leader of that cluster. Since every cluster is associated with a predefined area in the grid, if the density and distribution of the nodes change a new grid is needed and the clustering algorithm must be performed again in the whole topology.

One strong advantage of low mobility networks is that only few nodes need to be equipped with a GPS system; these will be used as static anchor nodes, the rest of the nodes can obtain their location by means of receiving beacon frames from the anchor nodes and by performing a localization mechanism. Several of these localization mechanisms were proposed where, Receive Signal Strength, different time of arrival of the beacon frames, angle of arrival or even the antenna directivity is used to calculate the position of a nodes. As an example [32], an algorithm proposed for WSNs can be applied to MANETs in low mobility scenarios. However, it must be taken into account that localization mechanisms use anchor nodes to assist in the localization of other nodes in the network and the accuracy of the mechanism highly decreases when the anchor nodes move.

6.4.2 Signal Environment-Based Solutions

The main limitation associated with GPS-based clustering algorithms is the energy cost. Other drawbacks are the limited achievable accuracy of this system which can introduce an uncertainty of several meters, and the difficulty of using it indoor environments. This section describes several proposals that do not require the availability of GPS systems. These algorithms are normally based in signal or environmental information, like received signal strength, changes in the frequency of the communication (Doppler effect), changes in the neighbouring nodes or probabilistic measurements of link availability.

Although several versions were proposed, the first signal strength based clustering algorithm is the well-known MOBIC [25]. MOBIC periodically broadcasts control messages for cluster formation and maintenance operations. These messages are also used to compute the relative mobility of a node with respect to each of its neighbours. The variability of the signal strength in two consecutive control messages is assessed using the following "Relative Mobility" formula for each node v with respect to a neighbor X_i:

$$RM_v(X_i) = 10 \log_{10} \frac{RSS_{\text{new}}(X_i)}{RSS_{\text{old}}(X_i)} \tag{6.9}$$

The "Mobility Prediction" of the node v is computed with the variance (relative to zero) of the RM with respect to all its neighbours:

$$MP_v = \text{var}_0(RM_v(X_1), RM_v(X_2), \ldots, RM_v(X_n)) \qquad (6.10)$$

where X_1 to X_m are the neighbouring nodes. The nodes with lower mobility in the neighbourhood are chosen to be *clusterheads*. Unfortunately, MOBIC authors do not provide any mechanism to mitigate the effect of fading and multipath that may lead to a misinterpretation of a node's mobility.

The Doppler effect is used in several proposals, like [33] in which the variations in the received frequency with respect to the expected transmission frequency can be used to assess how fast a neighbour is approaching or increasing its distance. The same problems identified in MOBIC are also present in this proposal. In general, the reliability of metrics based in signal measurements is subject to channel conditions.

To avoid the problem of unpredictable effects of channel conditions, in [34] the authors assessed the stability of a node and categorize a node by static or moving by measuring the change of neighbouring nodes and the rate of appearance of a neighbours in a given period of time. The authors proposed two metrics to assess stability of nodes based on the change of neighbours. Nevertheless, these metrics perform under the assumption that the network is sparse and nodes are mostly concentrated in hotspots; that is "moving" mobile nodes do not encounter more mobile "moving" neighbours than the "static" nodes in hotspots. Thus, the applicability of this algorithm is reduced to the scenarios where these strict characteristics apply. The main advantage is the complete independence from channel conditions and the operability of the algorithm without any kind of location information.

Another solution to adapt clusters to mobile environments is the DDCA algorithm [35] based on dynamically adapting the size of the cluster. The nodes join a cluster if there is at least a probability α that all cluster-members connect to it during a given time t. There is no limit in the number of hops, and a newcomer can connect directly to any member of the cluster. The fulfilment of this requirement is assessed when a node sends a request to join a cluster to a node that already belongs to that cluster. The main problem is that if the stability of a link in a connected node decreases then this (α, t) requirement could be compromised for the remaining nodes. Big clusters are formed when the network is static and stable. The cluster size is reduced when the conditions of the channel or the mobility of the nodes make the topology not stable.

6.5 Limitations in Mobile-Aware Clustering Algorithms for Urban Mobility

The application of clustering to urban scenarios it should be taken into account the characteristics of urban walking mobility and channel conditions. Although GPS systems do not face the problem of channel variations, GPS-based algorithms are not energy efficient. The assumption that handsets are equipped with GPS capabilities do not imply that these systems are activated in mobile terminals, and there is actually a high percentage of terminals that switch off this system due to battery constraints

However, clustering algorithms using channel measurements are affected by the strong multipath effects of urban scenarios. According the Friis formula for strong multipath effects the attenuation produced has an exponent n between 3 and 4, instead of 2 (considered in case of absence of fading and multipath):

$$\frac{P_r}{P_t} = G_r G_t \frac{\lambda}{4\pi R} \tag{6.11}$$

Thus the Relative Mobility computed by MOBIC would be

$$RM = 10\log_{10}\left(\frac{RSS_{new}}{RSS_{old}}\right) = 10\log_{10}\left(\frac{R_{old}}{R_{new}}\right)^n \tag{6.12}$$

where R is the direct distance between the nodes.

Other limitation of relative mobility based protocols (regardless if based on GPS or signal measurements) is the high re-affiliation rate. These algorithms select as *clusterheads* the nodes that have an "absolute" difference of speed relative to the neighbourhood lower than the rest. Clusters are connected with the majority of the neighbours for longer times. However, a small portion of nodes travelling faster or slower than the rest would leave and join new clusters frequently and for short periods of time. These re-affiliations produce a change in the cluster that forces the members to adapt the intra-cluster communication strategy and update intra-cluster information. To reduce the rate of affiliations, in [23] authors created GPS based heuristics introducing an entropy metric. Instead of choosing *clusterheads* with the lowest relative mobility they assess the relative mobilities as probabilities and compute the entropy of the distribution obtained. The node with the higher entropy is the one seeing the neighbourhood more ordered and should be the one elected as a *clusterhead*.

For every pair of nodes n and m they compute the following equations, to assess as probabilities the variation in the distance between the nodes in several consecutive samples:

$$\mathbf{p}(m, n, t) = \mathbf{p}(m, t) - \mathbf{p}(n, t) \qquad (6.13)$$

$$a_{m,n} = \frac{1}{N} \sum_{i=1}^{N} |\mathbf{p}(m, n, t)| \qquad (6.14)$$

Once a node m has computed all probabilities with respect to all neighbours it computes the entropy metric H_m, where F_m is the set of all neighbouring nodes of m and $C()$ is a function that counts the number of nodes in the set.

$$H_m(t, \Delta_t) = \frac{-\sum_{k \in F_m} P_k(t, \Delta_t) \log P_k(t, \Delta_t)}{\log C(F_m)} \qquad (6.15)$$

Figure 6.3 shows this difference in the election of the *clusterhead* for relative mobility and the described entropic metric (in this example we use the metric proposed by MBC algorithm described in Section 6.4.1.

This approach, however, cannot be applied in urban scenarios, since it may produce wrong elections when nodes move in opposite directions. As an example, a scenario where one node moves in one direction and finds several nodes moving in opposite direction would fail in cluster formation phase. At the time of cluster formation this node would have the neighbourhood more ordered than any of the other nodes walking in group in opposite direction (it has a uniform distribution of the neighbouring group of nodes moving directly towards him). This would lead to the election of the node that would be the first one leaving the group. Figure 6.4 depicts this problem.

The problem derived from the double direction feature is present in many proposals. A typical urban scenario such as people walking in the streets can be divided in two possible directions, and current clustering algorithms do not provide techniques to separate these two directions at the time of cluster formation. Even if group mobility is taken into consideration, there are no mechanisms limiting nodes from joining groups of nodes moving in opposite direction this situation highly increases the number of affiliations.

Figure 6.5 shows the results of a simulation where 90 nodes moving (walking speed between 1 and 2.5 meters per second) in horizontal lines in two possible (opposite) directions; any node can change direction at any time. The simulation records the time that every pair of nodes is in contact, and plots the number of links (Y axis) for a given time of link availability (X axis). We can appreciate a peak at the beginning showing that the majority of the links are established for a really short period of time. These links corresponds with nodes walking in opposite direction.

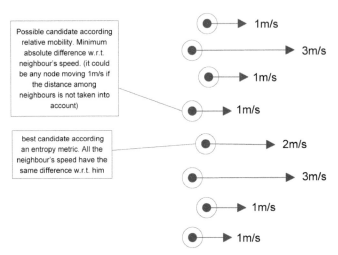

Figure 6.3 In this example the entropy metric elects a clusterhead that is more time in contact with the whole set of members. On the other hand, relative mobility chooses a clusterhead that is even more time in contact but only with a subset of the nodes in the cluster.

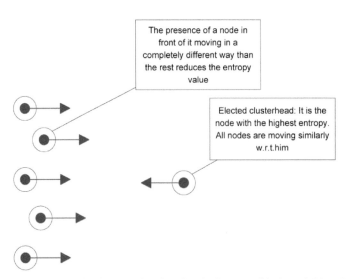

Figure 6.4 The node moving in opposite direction is the one with the neighbourhood more ordered. Thus, it is elected as clusterhead.

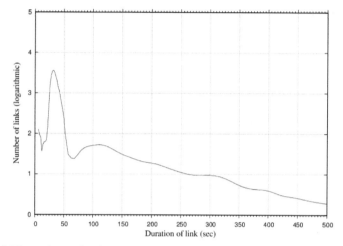

Figure 6.5 The node moving in opposite direction is the one with the neighbourhood more ordered. Thus, it is elected as clusterhead.

In conclusion, mobility-aware algorithms face problems of energy consumption in case of using GPS or are subject to channel conditions if they use signal related information (RSS, frequency, etc.). In case of GPS-based algorithms we could limit the usage of this system to only a few nodes but these should be static to be used as anchors, otherwise the accuracy decreases. Centralized solutions, that provide higher processing capabilities and data structures to gather and assess nodes mobility with low periodicity, can be an improvement by means of reducing the times that a node need to obtain its location and send this information [34]. Nevertheless, MANETs are expected to be deployed in infrastructure-less scenarios, this centralization cannot be accomplished. Other solutions that do not use mobility information do not face such problems, but are limited in their scope for scenarios with specific characteristics where some assumptions about the mobility, distribution and density of the nodes must hold.

6.6 New Trends in Clustering Techniques

New proposals in clustering algorithms are starting to take into account the diversity in hardware capabilities of the different mobile terminals. Yoo and Park [2] used Bluetooth for short range communication due to its low cost communication in terms of energy, and Wi-Fi for increased range (and data rate) between the *clusterhead* and the Access Point. In such strategy a *cluster-head* with two wireless interfaces is needed, and this requirement must be included in the algorithm for *clusterhead* election. Similarly, the processing capabilities, energy pattern consumption, memory or any kind of hardware features can be included in the heuristic to choose *clusterheads*.

The set of tasks performed by the *clusterhead* of a cluster is also an open discussion. Although some proposals describe protocols where *cluster-heads* do not have any special task, and all nodes are considered identical for the topology, they still use *clusterheads* to form the clusters [6]. In this case the *clusterhead* is the closest node to all the remaining members and directly connected in case of 1-hop clusters. This connectivity makes the *clusterhead* the best candidate to rule the cluster coordination – not taking advantage of this feature is a great disadvantage. Although many tasks can be shared, an example of sharing long range transmission task is given in [36], coordination should be always performed by the leader of the cluster. Related with task sharing there is ongoing research for task allocation algorithms. In [37] a cooperative method is proposed where the *clusterhead* selects sets of nodes to execute the different types of tasks, and where a voting system is implemented.

Multi-user diversity, where the different channel conditions of different users are taken into account before transmission, is also an example of how to use coordination in the *clusterhead* while sharing tasks. As an example, in WSNs the *clusterhead* is usually in charge of sending data to the base station in long range communication, but a more efficient deployment would give the chance to the *clusterhead* to choose dynamically among the members of the cluster the one with best channel conditions to transmit in long range. Channel diversity, where nodes constantly take into account the conditions of the channel allows as well efficient strategies when data has not strict delay constraints. The *clusterhead* faces the decision of "who" transmits in long range and "when" to use the best channel conditions and additionally save energy in long range communication.

Other problem related with clustering techniques that is being actively researched is channel allocation. The main problem is not actually how to deploy channels in clusters but how to choose the common control channel for control message exchange in cluster formation and maintenance. In urban scenarios many unlicensed wireless channels are already taken, and the channels available in one area can be crowded in the next location. Cognitive radio is typically the proposed approach towards such techniques. In [38] the authors used cognitive radio with a nature inspired algorithm based in swarm intelligence to choose a common channel among the nodes. Although they define a cluster as the group of nodes under the same channel, with no-limits in size or any control in the topology, this technique can be used to choose the control channel used by several clusters. Anyhow, clusters must be aware about the possible changes in the control channel in different areas.

6.7 Considerations and Future Directions

Many simulators have been developed, commercial and free software, and even handmade simulators, constructed from the scratch. Avoiding the use of complex simulators is a good solution when the computation of the results is not dependant on the simulation of multiple environment parameters, and needs only the computation of a smaller subset of variables in a limited number of layers. Although home-made simulators can produce the correct results, there is not common-ground for results comparison and research community would have problems in giving the appropriate recognition when compared to other studies that use well-known and community-accepted simulators. In [39] the authors presented a study of most common simulation tools used in reputed conferences and high impact factor journals. According to this study, ns2 and Matlab are the most well accepted simulators followed by Opnet. ns2 has a clustering framework [40] where several algorithms are already implemented and can be modified.

Apart from the simulator, researchers also need to generate the scenario. Among the free tools available the most known are IMPORTANT framework [41] and BonnMotion [42]. Both of them are capable to output the scenario in a ns2 file format. Scenario generators include several models that need proper selection to adapt the mobility of the nodes in the simulation to the desired scenario. Random waypoint model was initially the most used although it is not the more realistic - nodes move in a square, choosing randomly the destination point and the speed. Every time a node gets the destination point it chooses another one till the end of the simulation. In rare cases a real ad hoc scenario is described like in random waypoint model, and other models have been suggested for ad hoc mobility. In RPGM (Reference Point Group Mobility) [43] model nodes move in groups and a common destination point is given to each group. This model is usually applied for battlefield scenarios where soldiers form troops moving together. However for urban mobility a square with complete freedom does not depict the typical scenario, where users walk through bounded walking paths. Manhattan model was designed for this kind of scenarios, where nodes move in streets continuously in the two possible senses and there is a probability of changing street or continue in the current one. Modifications on this model were done to include pauses simulating traffic lights, and also vehicular together with pedestrian mobility.

Acknowledgements

The research leading to these results has received funding from the European Community's Seventh Framework Programme (FP7/2007-2013) under grant agreement No. 264759 (GREENET). Victor Sucasas and Hugo Marques are PhD students at the University of Surrey.

References

[1] C. E. Perkins, *Ad Hoc Networking*, Adison Wesley, 2001.

[2] J.-W. Yoo and K. H. Park, A cooperative clustering protocol for energy saving of mobile devices with WLAN and Bluetooth interfaces, *IEEE Transactions on Mobile Computing*, vol. 10, no. 4, pp. 491–504, April 2011. DOI: 10.1109/TMC.2010.161.

[3] C2power project, http://www.ict-c2power.eu, accessed: 30/06/2012.

[4] G. V. Zaia, M. Guarnera, M. Villari, A. Zaia, and A. Puliafito, Manet: Possible applications with PDF in wireless imaging environment, in *Proceedings of 13th IEEE International Symposium on Personal, Indoor and Mobile Radio Communications*, pp. 2394–2398, 2002.

[5] X. Hong, K. Xu, and M. Gerla, Scalable routing protocols for mobile ad hoc networks, *Network, IEEE*, vol. 16, no. 4, pp. 11–21, August 2002. DOI: 10.1109/MNET.2002.1020231.

[6] C. R. Lin and M. Gerla, Adaptive clustering for mobile wireless networks, *IEEE Journal on Selected Areas in Communications*, vol. 15, no. 7, pp. 1265–1275, September 1997. DOI: 10.1109/49.622910.

[7] R. P. A. M. L. Sichitiu, Angle of arrival localization for wireless sensor networks, in *Proceedings of IEEE Communications Society Conference on Sensor, Mesh and Ad Hoc Communications and Networks*, 2006.

[8] A. Savvides, C. Chieh Han, and M. B. Srivastava, Dynamic fine-grained localization in ad-hoc networks of sensors, in *Proceedings of the Seventh Annual International Conference on Mobile Computing and Networking (Mobicom 2001)*, pp. 166–179, 2001.

[9] J. Yu and P. Chong, A survey of clustering schemes for mobile ad hoc networks, *IEEE Communications Surveys Tutorials*, vol. 7, no. 1, pp. 32–48, 2005.

[10] J. Yu and P. Chong, 3HBAC (3-hop between adjacent clusterheads): A novel non-overlapping clustering algorithm for mobile ad hoc networks, in *Proceedings of IEEE Pacific Rim Conference on Communications, Computers and Signal Processing (PACRIM 2003)*, vol. 1, pp. 318–321, August 2003.

[11] H. P. Frank and F. Katz, *Cooperation in Wireless Networks: Principles and Applications: Real Egoistic Behavior Is to Cooperate!*. Springer-Verlag, New York, 2006.

[12] M. Chatterjee, S. Sas, and D. Turgut, An on-demand weighted clustering algorithm (WCA) for ad hoc networks, in *Proceedings of IEEE Conference on Global Telecommunications (GLOBECOM'00)*, vol. 3, pp. 1697 –1701. IEEE, 2000.

[13] D. Baker and A. Ephremides, The architectural organization of a mobile radio network via a distributed algorithm, *IEEE Transactions on Communications*, vol. 29, no. 11, pp. 1694–1701, November 1981.

[14] M. Gerla and J. T. Chieh Tsai, Multicluster, mobile, multimedia radio network, *Journal of Wireless Networks*, vol. 1, pp. 255–265, 1995.

[15] J. Wu and H. Li, On calculating connected dominating set for efficient routing in ad hoc wireless networks, in *Proceedings of the 3rd International Workshop on Discrete Algorithms and Methods for Mobile Computing and Communications*, ser. DIALM '99, pp. 7–14. ACM, New York, 1999. [Online] Available: http://doi.acm.org/10.1145/313239.313261.

[16] Y. P. Chen and A. L. Liestman, Approximating minimum size weakly-connected dominating sets for clustering mobile ad hoc networks, in *Proceedings of the 3rd ACM International Symposium on Mobile Ad Hoc Networking & Computing*, ser. MobiHoc '02, pp. 165–172. ACM, New York 2002. [Online] Available: http://doi.acm.org/10.1145/513800.513821.

[17] C. Chuan Chiang, M. Gerla, and L. Zhang, Forwarding group multicast protocol (FGMP) for multihop, mobile wireless networks, 1998.

[18] T. J. Kwon, M. Gerla, V. Varma, M. Barton, and T. Hsing, Efficient flooding with passive clustering – An overhead-free selective forward mechanism for ad hoc/sensor networks, *Proceedings of the IEEE*, vol. 91, no. 8, pp. 1210–1220, August 2003.

[19] W. R. Heinzelman, A. Chandrakasan, and H. Balakrishnan, Energy-efficient communication protocol for wireless microsensor networks, in *Proceedings of the 33rd Hawaii International Conference on System Sciences*, vol. 8, ser. HICSS '00, p. 8020. IEEE Computer Society, Washington, DC, 2000. [Online] Available: http://dl.acm.org/citation.cfm?id=820264.820485.

[20] O. Younis and S. Fahmy, Heed: A hybrid, energy-efficient, distributed clustering approach for ad hoc sensor networks, *IEEE Transactions on Mobile Computing*, vol. 3, no. 4, pp. 366–379, October/December 2004.

[21] A. Amis and R. Prakash, Load-balancing clusters in wireless ad hoc networks, in *Proceedings of 3rd IEEE Symposium on Application-Specific Systems and Software Engineering Technology*, pp. 25 –32, IEEE, 2000.

[22] T. Ohta, S. Inoue, and Y. Kakuda, An adaptive multihop clustering scheme for highly mobile ad hoc networks, in *Proceedings of the Sixth International Symposium on Autonomous Decentralized Systems (ISADS2003)*, pp. 293–300, April 2003.

[23] Y.-X. Wang and F. Bao, An entropy-based weighted clustering algorithm and its optimization for ad hoc networks, in *Proceedings of the Third IEEE International Conference on Wireless and Mobile Computing, Networking and Communications (WiMOB2007)*, p. 56, October 2007.

[24] H. Cheng, J. Cao, X. Wang, S. K. Das, and S. Yang, Stability-aware multi-metric clustering in mobile ad hoc networks with group mobility, *Wirel. Commun. Mob. Comput.*, vol. 9, no. 6, pp. 759–771, June 2009. DOI: 10.1002/wcm.v9:6.

[25] P. Basu, N. Khan, and T. Little, A mobility based metric for clustering in mobile ad hoc networks, in *Proceedings of the International Conference on Distributed Computing Systems Workshop*, pp. 413–418, April 2001.

[26] Y. Zhang and J. M. Ng, A distributed group mobility adaptive clustering algorithm for mobile ad hoc networks, in *Proceedings of the IEEE International Conference on Communications (ICC'08)*, pp. 3161–3165, May 2008.

[27] K. Liu, J. Su, J. Zhang, F. Liu, and C. Gong, A novel stable cluster protocol for mobile ad hoc networks, in *IEEE International Symposium on Microwave, Antenna, Propagation and EMC Technologies for Wireless Communications (MAPE 2005)*, vol. 2, pp. 1328–1332, August 2005.

[28] B. An and S. Papavassiliou, A mobility-based clustering approach to support mobility management and multicast routing in mobile ad-hoc wireless networks, *Int. J. Netw. Manag.*, vol. 11, no. 6, pp. 387–395, November 2001. DOI: 10.1002/nem.415.

[29] C. S. Jensen, D. Lin, S. Member, and B. C. Ooi, Continuous clustering of moving objects, *IEEE TKDE*, pp. 1161–1173, 2007.

[30] S. Leng, Y. Zhang, H.-H. Chen, L. Zhang, and K. Liu, A novel k-hop compound metric based clustering scheme for ad hoc wireless networks, *IEEE Transactions on Wireless Communications*, vol. 8, no. 1, pp. 367–375, January 2009.

[31] L. Jiancai and H. Xiao, A novel clustering algorithm based on gps of the mobile ad hoc network, in *Proceedings of the 5th International Conference on Wireless Communications, Networking and Mobile Computing (WiCom'09)*, pp. 1–4, September 2009.

[32] K. Ali, S. Neogy, and P. Das, Optimal energy-based clustering with GPS-enabled sensor nodes, in *Proceedings of the Fourth International Conference on Sensor Technologies and Applications (SENSORCOMM)*, pp. 13–18, July 2010.

[33] C. R. Z. D. Ni Minming, and Zhong Zhangdui, Doppler shift based stable clustering scheme for mobile ad hoc networks, 2012.

[34] B. Gu and X. Hong, Mobility identification and clustering in sparse mobile networks, in *Proceedings of IEEE Military Communications Conference (MILCOM 2009)*, pp. 1–7, October 2009.

[35] A. McDonald and T. Znati, Design and performance of a distributed dynamic clustering algorithm for ad-hoc networks, in *Proceedings of 34th Annual Simulation Symposium*, pp. 27–35, 2001.

[36] M. Alam and J. Rodriguez, A dual head clustering mechanism for energy efficient WSNs, in *Proceedings of MOBILIGHT*, pp. 380–387, 2010.

[37] Y. Yang, X. Song Qiu, L. Ming Meng, and L. Lan Rui, A self-adaptive method of task allocation in clustering-based manets, in *Proceedings of IEEE Network Operations and Management Symposium (NOMS)*, April 2010, pp. 440–447, IEEE, 2010.

[38] T. Chen, H. Zhang, and M. D. Katz, Cloud networking formation in cogmesh environment, *CoRR*, vol. abs/0904.2028, 2009. [Online] Available: http://dblp.uni-trier.de/db/journals/corr/corr0904.html#abs-0904-2028.

[39] N. I. Sarkar and S. A. Halim, A review of simulation of telecommunication networks: Simulators, classification, comparison, methodologies, and recommendations, *Journal of Selected Areas in Telecommunications (JSAT)*, March 2011.

[40] S. Basagni, M. Mastrogiovanni, A. Panconesi, and C. Petrioli, Localized protocols for ad hoc clustering and backbone formation: A performance comparison, *IEEE Transactions on Parallel and Distributed Systems*, vol. 17, no. 4, pp. 292–306, April 2006.

[41] F. Bai, N. Sadagopan, and A. Helmy, Important: A framework to systematically analyze the impact of mobility on performance of routing protocols for adhoc networks, in *Proceedings of Twenty-Second Annual Joint Conference of the IEEE Computer and Communications (INFOCOM 2003)*, vol. 2, March/April 2003, pp. 825–835. IEEE, 2003.

[42] N. Aschenbruck, R. Ernst, E. Gerhards-Padilla, and M. Schwamborn, Bonnmotion: A mobility scenario generation and analysis tool, in *Proceedings of the 3rd International ICST Conference on Simulation Tools and Techniques*, ser. SIMUTools '10, pp. 51:1–51:10. ICST (Institute for Computer Sciences, Social-Informatics and Telecommunications Engineering), Brussels, Belgium, 2010. [Online] Available: http://dx.doi.org/10.4108/ICST.SIMUTOOLS2010.8684.

[43] X. Hong, M. Gerla, G. Pei, and C.-C. Chiang, A group mobility model for ad hoc wireless networks, in *Proceedings of the 2nd ACM international workshop on Modeling, Analysis and Simulation of Wireless and Mobile Systems*, ser. MSWiM '99, pp. 53–60. ACM, New York, 1999. [Online] Available: http://doi.acm.org/10.1145/313237.313248.

7

Network Coding for Wireless Networking

Riccardo Bassoli[1,2], Hugo Marques[2,3], Jonathan Rodriguez[1], and Rahim Tafazolli[2]

[1]*Instituto de Telecomunições, Aveiro, Portugal*
[2]*CCSR, University of Surrey, UK*
[3]*Instituto Politécnico de Castelo Branco, Portugal*
e-mail: bassoli@av.it.pt

Abstract

Network coding is a promising technique to increase the performances of wireless networks in terms of throughput, energy efficiency and complexity. In fact, coding instead of routing can increase the transmission rate in multicast scenarios and simplify the design of protocols and algorithms. Moreover, some initial results of network coding in wireless networks have shown its capability to achieve minimum energy in multicast scenarios compared to 'classical' routing protocols. This chapter presents the theoretical framework for wireless network coding applications. Network coding research is increasingly gathering pace since it has been shown to be a powerful solution for energy efficient networks and video streaming applications. However, there are still open issues on how many of these solutions can be actually implemented in practice. This chapter investigates some of the most typical network coding solutions for implementation on real networks and highlights the challenges that lie ahead.

Keywords: Network coding, wireless networks, practical network coding, subspace codes.

Shahid Mumtaz and Jonathan Rodriguez (Eds.), Green Communication in 4G Wireless Systems, 117–138.

7.1 Introduction

In 1948, the discipline called information theory had its inception thanks to the fundamental work of Shannon [1]. From that moment, the information transmitted through a network had been interpreted as a commodity so that the only capability of terminals was to store and to forward. A few years later, Ford and Fulkerson [2] and Elias et al. [3] contemporaneously and independently demonstrated the max-flow min-cut theorem, which defined the possible maximum flow achievable in a graph by transmitting information as a commodity. In 2000, the seminal article [4] demonstrated the max-flow min-cut theorem for information flow. By considering information flow and not commodity flow, the nodes are allowed to manipulate the information received and to perform some coding operations: this intuition represented the beginning of network coding. Network coding is a kind of coding operation executed at packet level. Prior to that, coding theory only knew two kind of operations: source coding, the way to compress the information at the source to increase the efficiency in the transmission, and channel coding, the operation of introducing redundant bits in the information sequence to protect it against errors due to the noisy channel. Thus, the principal idea of network coding is that a node can transmit functions of the messages received on the ingoing edges onto the outgoing ones. Next, Li et al. [5] demonstrated that linear network codes for multicast were able to achieve the max-flow min-cut bound and the authors provided a construction for static networks. Then, Koetter and Médard [6] described an algebraic geometric framework to study and design network coding for multicast. This different approach allowed the use of the several tools available in algebraic geometry to improve network codes. Moreover, the link found between this result and Edmonds' theorem opened the way to the development of random linear network codes (RLNC) [7]. By randomly choosing the coefficients of the linear combinations over a finite field, RLNC permitted an efficient and easier implementation of network coding in wireless and dynamic scenarios. In 2003, Chou et al. [8] described the first implementation of network coding by using RLNC for a wired internet network. They analyzed the issue of transmitting coding vectors into the header of the packets to make the decoding process possible for the sinks. In 2008, Koetter and Kschischang [9] designed an innovative kind of network codes, called subspace codes, that promised to avoid the transmission of the coding vectors to the destination. The main idea was to consider no knowledge on the coding coefficients but only on the vector space spanned by the source packets: because of that, these network codes

Table 7.1 Description of the average energy of random multicast connections of unit rate for different sizes of the random wireless network. The energy to transmit at unit rate to a distance d is considered d^2 [19].

Network size	Approach	Complexity of deterministic algorithms			
		2 sinks	4 sinks	8 sinks	16 sinks
20 nodes	MIP algorithm	30.6	33.8	41.6	47.4
	Network coding	15.5	23.3	29.9	38.1
30 nodes	MIP algorithm	26.8	31.9	37.7	43.3
	Network coding	15.4	21.7	28.3	37.8
40 nodes	MIP algorithm	24.4	29.3	35.1	42.3
	Network coding	14.5	20.6	25.6	30.5
50 nodes	MIP algorithm	22.6	27.3	32.8	37.3
	Network coding	12.8	17.7	25.3	30.3

were named noncoherent. So, by transmitting packets without coding vectors, it became possible to avoid the overhead in packet headers. Nevertheless, the main drawback was the complexity of the implementation, especially of the decoding algorithms. Then, Silva and coworkers [10, 11] developed a simpler instance of subspace codes by taking advantage of a theoretical connection between these codes and rank-metric codes. In this way it was possible to reduce the complexity of the initial framework. The research in subspace codes for network coding resulted in four different kinds of subspace codes: lifted rank-metric codes [10, 11], padded codes [12–14], lifted Ferrers diagram codes [15] and codes obtained by linear programming [16]. In 2009, Jafari et al. [17] proposed a different way to treat the overhead given by the coding vectors: the authors provided an algorithm for the compression of the coding vectors making their transmission in the packets more efficient. The importance of keeping the coding vectors was the discovery of their utility in passively inferring information on network topology.

The theoretical results above were fundamentals for the definition of a network coding framework useful in wireless scenarios. The need for this framework is given by the fact that network coding can be better than 'classical' routing protocols not only to increase the performance in terms of rate and throughput, but even to simplify the design of the network protocols and to increase the energy efficiency. These last considerations are particularly important in wireless networks where the battery life of the wireless nodes and the complexity of the algorithms are critical parameters. In 2005, Lun

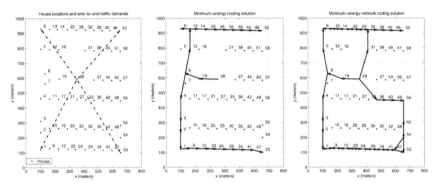

Figure 7.1 The image on the left shows the example in which node 28 is the source transmitting to the set of sinks 1, 6, 52, 53. In the centre the connectivity graph of the solution obtained for routing is shown and on the right, the solution calculated by using network coding [21].

et al. [18, 19] presented a method to achieve minimum-cost multicast with RLNC in lossless wireless networks. In particular, this article described two decentralized polynomial-time algorithms to compute minimum-cost sub-graphs by using linear cost functions and strictly convex cost functions. These approaches are attractive for overlay and multi-hop wireless networks. The initial NP-complete computational complexity of this problem suggested the authors to employ a heuristic technique, called multicast incremental power (MIP) algorithm [20]. Table 7.1 reports the results of the simulations in [19]: they provided that network coding reduced the range in the average total energy of multicast problem in random wireless networks of different size from 13 to 49%, compared with classical routing solution with MIP algorithm. In the same year, Wu et al. [21] demonstrated that RLNC can reach the minimum energy-per-bit in a wireless multicast ad hoc network. Moreover, the authors found that, by applying network coding instead of routing, the minimum energy-per-bit is achieved in polynomial time (the construction of a minimum-energy multicast tree by using 'classical' routing is an NP-hard problem). In Figure 7.1, the results of their simulations appear: the considered scenario is a community wireless network consisting of 58 houses. The authors' simulations confirmed that the minimum energy-per-bit with network coding was 98.08% of the one calculated with routing and the computation of the optimal network coding solution took less than 1/30 of the time to compute the optimal routing solution.

The remainder of this chapter is a tutorial on the theoretical model to analyze network coding in wireless networks. Following a review of the fun-

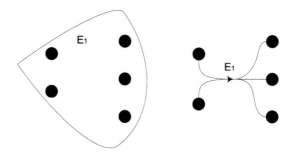

Figure 7.2 Illustration of an undirected hyperedge (left) with two start nodes and three end nodes. Representation of a directed hyperedge (right).

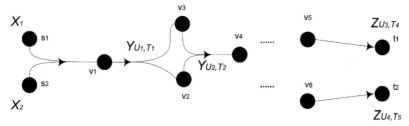

Figure 7.3 Example of two parts of a wireless network modeled as a hypergraph – the beginning and the end. There are two sources and two sinks.

damentals of RLNC theory in the wireless domain, we describe the main challenges that arise when implementing these ideas on real networks. We focus here on novel approaches through subspace codes and compressed coding vectors, mainly proposed to solve the problem of the overhead in wireless network coding. The discussion touches generic theoretical concepts and avoids detailed mathematical analysis of encoding and decoding algorithms, left for further studies.

7.2 A Generalized Model for Network Coding in Wireless Scenarios: Random Linear Network Coding in Hypergraphs

The following introduction to RLNC takes into account the theoretic case of delay-free and lossless wireless networks, which are modelled as directed hypergraphs. We will start, firstly, with some notions about hypergraph theory. Algebraic network coding theory is also discussed in [22].

A hypergraph is a pair $\mathcal{H} = (V, E)$, where $V = \{V_1, \ldots, V_n\}$ is the set of nodes and $E = \{E_1, \ldots, E_m\}$ is the set of hyperedges, with E_i subsets of V. A hypergraph represents a generalization of the classical concept of graph: in fact, if $|E_i| = 2$ for each E_i, the hyperedges are normal edges and the hypergraph becomes a graph. Next, an hyperarc is a directed hyperedge $E = (U, T)$, where U, T are disjoint subsets of vertices: U is the set of start nodes and T is the set of end nodes. A directed hypergraph is a hypergraph with directed hyperedges. Figure 7.2 shows an example of undirected hyperedge on the left and the respective directed on the right. By considering the most general case with arbitrary coding functions, it is possible to define three kind of random processes, transmitted on hyperarcs:

- X_i, the random source process sent by the source i;
- Y_{U_l, T_k}, the random processes transmitted by the set of nodes U_l to the set of nodes T_l;
- Z_{U_j, T_k}, the random processes received by the sinks from the set of nodes U_j.

In order to clarify these concepts, Figure 7.3 depicts an example of flow of random processes in a part of a hypergraph. The general mathematical equation which describes random processes on the hyperarcs, in case of scalar linear network coding:

$$Y_{U_l, T_p} = \sum_{U_l \subseteq T_k'} \beta_{(U_{j'}, T_k'),(U_l, T_p)} Y_{U_{j'}, T_k'} + \begin{cases} \sum_i \alpha_{i,(U_l, T_p)} X_i & \text{ifs} \in U_j' \\ 0 & \text{otherwise} \end{cases}$$

(7.1)

where $\alpha_{i,(U_l, T_p)}$, $\beta_{(U_{j'}, T_k'),(U_l, T_p)}$ are scalar elements in the finite field $\mathbb{F}_q$, called local coding coefficients. On the other hand, the random processes received by the sinks are described as follows:

$$Z_{U_j, T_k} = \sum_{U_j \subseteq T_p'} \varepsilon_{(U_j, T_k),(U_{l'}, T_p')} Y_{U_{l'}, T_p'}$$

(7.2)

where the scalar elements $\varepsilon_{(U_j, T_k),(U_{l'}, T_p')}$ are the decoding coefficients. The sinks need to receive a number of linearly independent random processes equal to the number of source processes to be able to decode the information sent by the sources. The scalar linear operations due to equation (7.1) can be interpreted as scalar linear functions dependent only from the source processes X_i. Hence, the random processes on hyperarcs can be rewritten into the form

$$Y_{U_l,T_p} = \sum_{i=1}^{s} c_{i,(U_l,T_p)} X_i \qquad (7.3)$$

where s is the number of sources and $c_{i,(U_l,T_p)}$ are the elements of the global coding vector on a hyperarc. By considering $\mathbf{x}$ as the vector of source random processes and $\mathbf{z}$ the vector of random processes received by the sinks, it is possible to define a transfer matrix $\mathbf{M}$, which describes the linear network code of the network and is a function of coefficients $\{\alpha_{i,j}, \beta_{i,j}, \varepsilon_{i,j}\}$. The matrix $\mathbf{M}$ is determined as the matrix product

$$\mathbf{z} = \mathbf{Mx}. \qquad (7.4)$$

Let us define the matrices $\mathbf{A}, \mathbf{B}, \mathbf{F}$ as

- $\mathbf{A} = (\alpha_{i,j})$, a $s \times |E|$ matrix, whose nonzero elements are the coefficients linearly combined with the source random processes;
- $\mathbf{F} = (\beta_{i,j})$, a $|E| \times |E|$ matrix, whose nonzero elements are the coefficients linearly combined with the random processes on the hyperarcs;
- $\mathbf{B} = (\varepsilon_{i,j})$, a $s \times |E|$ matrix, whose nonzero elements are the coefficients, that are combined with random processes on the incoming hyperarcs by the sink to form the output processes Z.

Hence, the transfer matrix in equation (7.4) can be written in the more explicit form

$$\mathbf{M} = \mathbf{A}(\mathbf{I} - \mathbf{F})^{-1}\mathbf{B}^{\mathrm{T}}. \qquad (7.5)$$

The description of $\mathbf{M}$ through the matrices of coefficients contributes to further clarify its structure. Specially, matrices $\mathbf{A}, \mathbf{B}$ have less influence over $\mathbf{M}$ than matrix $\mathbf{F}$, because $\mathbf{A}, \mathbf{B}$ are only related with linear combinations of sources and sinks random processes.

The design of a network code for a multicast scenario is called a network coding problem. Then, a network coding problem is called solvable if there exists for each sink node a flow of rate r between the source and each sink. This proposition, in the network coding algebraic framework above, is translated into

$$\det(\mathbf{M}) \neq 0 \qquad (7.6)$$

over the ring $\mathbb{F}_q[\ldots, \alpha_{i,j}, \ldots, \beta_{i,j}, \ldots, \varepsilon_{i,j}, \ldots]$. Constraint (7.6) guarantees that matrix $\mathbf{M}$ is invertible, permitting the sinks to find the solution of the linear system once they have collected all the random processes Z needed.

A random linear network code is a network code in which the elements of matrices $\mathbf{A}, \mathbf{F}$ are chosen randomly and independently from a finite field $\mathbb{F}_q$.

The main benefits of the randomized approach are the application in changing topologies, large networks or in the presence of dynamically varying
connections. In fact, in such scenarios, it is preferable the distributed nature
of random network coding instead of managing a difficult centralized coding
solution. Nevertheless, the fact of choosing randomly the coefficients of the
linear combinations has the drawback of introducing a probability to allow
an existing solution of the network coding problem. Hence, RLNC achieves
the multicast capacity with an exponential probability of value

$$\left(1 - \frac{d}{q}\right)^{\gamma} \tag{7.7}$$

which is approaching 1 by increasing the size of the alphabet: the variable d
represents the number of sinks and the variable γ is the number of hyperarcs,
which have associated random coefficients $(\alpha_{i,j}, \beta_{i,j})$.

7.3 Network Coding in Practical Scenarios

The previous section considered two random processes X, Y flowing through
the network. By contextualizing them in a real scenario, X, Y become bit-
streams, then divided into transmission packets. The packets are collections
of arrays of n bits $[\mathbf{x}_1, \ldots, \mathbf{x}_h]^{\mathrm{T}}$ and $[\mathbf{y}_1, \ldots, \mathbf{y}_h]^{\mathrm{T}}$ – represented by vectors
of n elements – which are interpreted as symbols over a finite field $\mathbb{F}_q$,
with $q = 2^n$. However, the application of network coding in a real wireless scenario needs to introduce some additional concepts. In fact, we have
to think about information flowing asynchronously, which can suffer delays
and losses, and eventually experience the possibility of congestion and link
failures. So, to be practical, the framework of RLNC should be designed with
a special packet structure taking in consideration new key characteristics to
overcome real network issues.

In order to make the real scenario synchronized, the packets are organized
into generations. A generation is a set of size h (generation size) of packets
sent by the same source, accomplished by labelling packets headers with a
generation number. By using this model, the packet transmitted by a node is
a linear combination of the packets – with the same generation number – it
received and the coefficients of the linear combinations are chosen according
to a uniform distribution over a finite field. Nevertheless, the generations
implied an extra cost in the packets: in fact, the operation of packet tagging

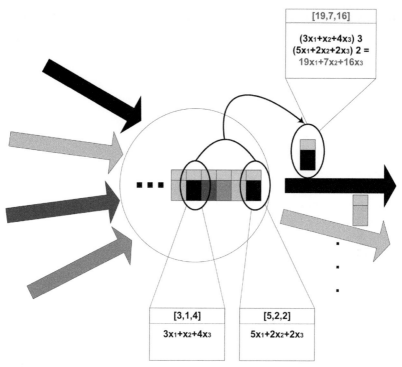

Figure 7.4 Representation of the operations at an intermediate node in a network using network coding. The buffer collects the packet from different generations (different gray shades) and the linear combinations on the outgoing channels are made among packets belonging to the same generation. The image also shows the presence in the header of the linear coefficients.

introduced h extra symbols in each packet. For example, if $h = 60$ and the wireless packet is 1500 bytes, the overhead is $60/15000 \approx 4\%$.

In the buffer of a network coding node the packets are not only sorted by generations but they are also classified into innovative and non-innovative packets. The former are new packets transporting vectors which increase the dimension of the subspace, spanned by the ones which are already in the buffer of the node; the latter are packets containing redundant information which do not increase the size of the subspace spanned by the packets already at the node so they can be deleted without loss of information. Figure 7.4 presents an example on these concepts.

On the receiver side, the sinks need to receive h packets belonging to the same generation to be able to decode the information of the source through Gaussian elimination. What a receiver knows is only the overall linear combination from the source to the receiver: this transformation is the global encoding vector **G**. This concept can be summarized by the following formula:

$$
\begin{bmatrix} \mathbf{y}_1 \\ \vdots \\ \mathbf{y}_h \end{bmatrix} = \begin{bmatrix} \mathbf{g}_1^{\mathrm{T}} & \cdots & \mathbf{g}_h^{\mathrm{T}} \end{bmatrix} \begin{bmatrix} \mathbf{x}_1 \\ \vdots \\ \mathbf{x}_h \end{bmatrix} = \mathbf{G} \begin{bmatrix} \mathbf{x}_1 \\ \vdots \\ \mathbf{x}_h \end{bmatrix} \tag{7.8}
$$

Let us consider the source packets indexed by integers $i = 1, \ldots, h$. The way to convey the coefficients of the linear combinations to the sinks is to add a vector with a nonzero element in the ith position. All the packets created later through linear combinations have its own coding vector of length h attached. This concept can be expressed by the following relation:

$$
\begin{bmatrix} \mathbf{g}_1^{\mathrm{T}} & \cdots & \mathbf{g}_h^{\mathrm{T}} & \begin{matrix} \mathbf{y}_1 \\ \vdots \\ \mathbf{y}_h \end{matrix} \end{bmatrix} = \begin{bmatrix} \mathbf{g}_1^{\mathrm{T}} & \cdots & \mathbf{g}_h^{\mathrm{T}} \end{bmatrix} \begin{bmatrix} \mathbf{I} & \begin{matrix} \mathbf{x}_1 \\ \vdots \\ \mathbf{x}_h \end{matrix} \end{bmatrix}. \tag{7.9}
$$

Then, the length of the coding vector in packets is $h \log q$. By using long packets the overhead given by coding vectors is small but for shorts packets (like in wireless communications), this overhead can be significant. So, the way to reduce it is to reduce the size of the field or the size of a generation. But the reduction of the field size increases the error probability of the random code and increase the possibility of retransmissions. So, the values of h and q have to be chosen according to the application and the kind of network. At the end, when the receiver has collected all the linearly independent packets of a generation it can decode the information.

7.4 Subspace Codes

As described in the previous section, it is clear that coding vectors in packet headers represent an unavoidable overhead to implement RLNC in real networks, with the effect of significantly augmenting the packet length when the transmission occurs in wireless networks (using short packets). The solution proposed to avoid this issue is the application of subspace coding. By assuming that the matrix **G** of the global encoding vectors is a random matrix, we state that the receiver has no knowledge about **G**. By defining sequence $[\mathbf{y}_1, \ldots, \mathbf{y}_h]^{\mathrm{T}}$ received by the sink as in equation (7.8), it is possible to prove that the characteristic that remains fixed on the receiver side is the row space spanned by the vectors/packets $[\mathbf{x}_1, \ldots, \mathbf{x}_h]^{\mathsf{T}}$ sent by the source. Due to this assumption that the sink has no information about the matrix $\mathbf{G}$ but only about the vector space spanned by the source packets, this different approach to network coding is called noncoherent. The theory of subspace codes is also discussed in [23].

In order to define a subspace code, we firstly consider a vector space $\mathbb{F}_q^n$, its Grassmannian $\mathcal{G}(k, n)$ and two subspaces $Q, W \subset \mathbb{F}_q^n$. The metric is the fundamental concept in coding theory for the definition of a code. Therefore, we start by defining d as the function $d : \mathcal{G}(k, n) \times \mathcal{G}(k, n) \to \mathbb{Z}_+$ such that

$$d(Q, W) := \dim(Q + W) - \dim(Q \cap W) \qquad (7.10)$$

Then, a code C is a nonempty subset of $\mathcal{G}(k, n)$. The minimum distance of this subspace code is defined as

$$d_{\min} := \min_{Q, W \in C : Q \neq W} d(Q, W). \qquad (7.11)$$

The initial subspace framework for network coding had the drawback of being difficult for applications in practice. Therefore, the description above was reformulated to translate codes in Grassmannians into codes using a rank-metric. In this way, it was possible to get several benefits – especially on the decoding side – provided by all the existing tools in coding theory for rank-metric codes. Before the description of the connection between subspace codes and rank-metric codes, it is useful to show some basic concepts on rank-metric codes.

A linear array code (or matrix code) $C(n \times m, k, d_{\min})$ is defined as a k-dimensional linear subspace of $\mathbb{F}_q^{n \times m}$, with the minimum weight $d_{\min}$ of any nonzero matrix in C, called the minimum distance of the code. Let $\mathbf{A}, \mathbf{B}$ be matrices in $\mathbb{F}_q^{n \times m}$, then the rank distance between them is defined as

$$d_R(\mathbf{A}, \mathbf{B}) := \mathrm{rank}(\mathbf{B} - \mathbf{A}). \qquad (7.12)$$

The use of this metric implies the name of rank-metric codes for matrix codes. Hence, from equation (7.11) the minimum rank distance for a rank-metric code is straightforward. The direct connection between subspace codes and rank-metric codes is given by the relation

$$\dim (\langle \mathbf{A} \rangle, \langle \mathbf{B} \rangle) = 2 d_R (\mathbf{A}_1, \mathbf{B}_1) \qquad (7.13)$$

where $\mathbf{A} = \begin{bmatrix} \mathbf{I} & \mathbf{A}_1 \end{bmatrix}$ and $\mathbf{B} = \begin{bmatrix} \mathbf{I} & \mathbf{B}_1 \end{bmatrix}$ are matrices such that $\mathbf{A}_1, \mathbf{B}_1 \in \mathbb{F}_q^{n \times m}$, and $\mathbf{I}$ is the identity matrix. We have shown the roots of noncoherent network coding, and provided a brief definition of its metric and its link with rank-metric codes. With this fundamental platform in place, we provide a brief introduction on the construction of some famous asymptotically good subspace codes.

The first family of subspace codes introduced is the one of lifted rank-metric codes. Let $\mathbf{A} \in \mathbb{F}_q^{n \times m}$ be a matrix and let $C \in \mathbb{F}_q^{n \times m}$ a code. Subspace $W(\mathbf{A})$ is called the lifting of the matrix $\mathbf{A}$ and it is defined as

$$W(\mathbf{A}) := \langle [\, \mathbf{I} \quad \mathbf{A} \,] \rangle \tag{7.14}$$

where $\mathbf{I}$ is an $n \times n$ identity matrix. It is straightforward the definition of a lifting $W(C)$ for a code C. Specially, $W(C)$ means the lifting of each code-word of the code and represents a constant dimension code. The expression of the distance metric for lifted rank-metric codes becomes

$$\begin{aligned} d(W(\mathbf{A}), W(\mathbf{B})) &= 2d_R(\mathbf{A}, \mathbf{B}) \\ d(W(C)) &= d_R(C) \end{aligned} \tag{7.15}$$

which preserves the same properties of rank-metric codes.

A second family of codes in noncoherent network coding is the one constituted by padded codes. Let $m, p \in \mathbb{Z}_+$, where $p < m$, and let $n = (m + 1)k + p$. Padded codes are codes obtained as a union of lifted product rank-metric codes, such that

$$P = \bigcup_{i=0}^{m-1} P_i \tag{7.16}$$

$$P_i = \langle [\, \mathbf{0}^1{}_{k \times k} \quad \cdots \quad \mathbf{0}^i{}_{k \times k} \quad \mathbf{I}_{k \times k} \quad \mathbf{c}^T_{i+1} \quad \cdots \quad \mathbf{c}^T_p \,] \rangle$$

with $\mathbf{c}_{i+1} \in C$ for $i = 0, \ldots, p - 2$ and $\mathbf{c}_p \in C'$, where $C \subseteq \mathbb{F}_q^{k \times k}$ and $C' \subseteq \mathbb{F}_q^{k \times (k+p)}$. If $C = C'$ the padded codes become the maximum rank distance (MRD) codes called Gabidulin codes. Moreover, if we also consider the codes having constant rank distance, padded codes become spread[1] codes. The main characteristics of spread codes are to have codewords, which are always subspaces of the same dimension and, given two different elements of the code only intersecting in the origin, these codes obtain the maximal possible distance among the subsets of the Grassmannian. So, a spread code S is a subset of $\mathcal{G}(\mathbb{F}_q^n, k)$ and is a q-ary code $\left[n, k, \left(\frac{q^n - 1}{q^k - 1}\right), 2k\right]$, which is defined according to relation (7.16).

---

[1] A spread is a partition of a vector space by subspaces of the same dimension and it is a subset of a finite Grassmannian $\mathcal{G}(k, n)$.

Lifted Ferrers diagram (FD) codes represent a generalization of subspace codes, constructed through a multilevel method from codes in projective spaces: in fact, FD rank-metric codes represents the main block in the multilevel construction in projective spaces. A code $C_{\mathcal{F}}$ is a $[\mathcal{F}, |C|, k]$ FD rank-metric code when the codewords are $n \times m$ matrices, in which all entries not in the $n \times m$ Ferrers diagram $\mathcal{F}$ are zeros: $|C|$ is the dimension of the code and k is the minimum rank distance of the code. The multilevel approach used to design these kinds of code can be summarized into four steps:

- Let us consider a binary constant-weight code $[n, m, 2k, l]$, where l is the weight. This initial code is called the skeleton code because the codewords are bases for subspaces in reduced echelon form. The characteristic of the skeleton is that it contains Ferrers diagram on which it is possible to build a rank-metric code.
- Next, it is necessary to calculate the echelon Ferrers form $EF(\mathbf{x})$ of the binary codeword $\mathbf{x} \in C$.
- Then, let us construct the $[\mathcal{F}, |C|, k]$ Ferrers diagram rank-metric code $C_{\mathcal{F}}$ for the Ferrers diagram $\mathcal{F}$ of $EF(\mathbf{x})$.
- The final step is to apply the lifting operation on $C_{\mathcal{F}}$. The result of this final step is a constant-dimension code which has echelon Ferrers form $EF(\mathbf{x})$.

It is important to note that the size of the obtained constant-dimension code depends on the choice of the skeleton code.

In the theory of subspace codes there is a family of constant-dimension codes, whose design is interpreted as an optimization problem in $\mathbb{Z}$. Let C be a constant-dimension code such that

$$d(Q, W) = 2k - 2\dim(Q \cap W) \qquad (7.17)$$

where k is the dimension of the code and $U, W \in C$. Because of the value of the minimum subspace distance that has to be even and less or equal to $2k$, equation (7.17) can be rewritten as

$$d(Q, W) \geq d. \qquad (7.18)$$

Thus, the design of these constant-dimension subspace codes becomes the problem of finding n subspaces of dimension k such that there is no subspace of dimension $k - d + 1$ in a pair of considered k-spaces in a projective space of order n. Formally, this can be translated saying that, given $\{Q_1, \ldots, Q_n\} \in \mathcal{G}(k, n)$ for all $i, j \in \{1, \ldots, n\}$, we can write

$$\dim(Q_i \cap Q_j) \leq k - d. \qquad (7.19)$$

By defining an incidence matrix **C** between the $(k-d+1)$-spaces labelling its rows and the k-spaces labelling its columns, we have a matrix with structure

$$M_{W,Q} := \begin{cases} 1 & W \subseteq Q \\ 0 & \text{otherwise} \end{cases} \tag{7.20}$$

Hence, given the vector $\mathbf{x} = \left(x_1, \ldots, x_{\left[\frac{n}{k}\right]} \right)$, the code construction follow the rule

$$\text{maximize} \sum_{1}^{\left[\frac{n}{k}\right]} x_i = m \quad \mathbf{Cx} \leq \begin{pmatrix} 1 \\ \vdots \\ 1 \end{pmatrix} \tag{7.21}$$

where m is the number of codewords of the constant-dimension code.

7.5 Compressed Coding Vectors

In this section we focus on some key results about coding vectors. The reason to move from 'classical' network coding to noncoherent network coding is to avoid the overhead caused by the necessary transmission of coding vectors inside packets. And besides being an issue, the coding vectors are also a passive source of information: in fact, the distribution of coding vectors in the network is not uniform but depends on the topology. Hence, this overhead can have a positive rule in passively inferring information about network topology such as the position of link or node failures and the presence of congestions. A possible alternative to the subspace and RLNC approaches described above, is to use compressed coding vectors to efficiently distribute the coding coefficients through the network.

Let us consider a code $[n, k, d_{\min}]$ with $d_{\min} = \min\{2r + 1, h + 1\}$, in which h are the packets in each generation and r are the packets – with $r < h$ – most needed by the receiver to determine the source packets linearly combined. The compression of coding vectors is possible if $r \ll h$, when the initial coding vectors become sparse. Next, let us consider the $p \times h$ parity check matrix **H** of this code, with $p = h - k$. The coding vectors assign to source packets $[\mathbf{x}_1, \ldots, \mathbf{x}_h]^T$ the columns of matrix **H**, such that the ith column is

$$\mathbf{h}_i = \mathbf{e}_i \mathbf{H}^T \tag{7.22}$$

where $\mathbf{e}_i$ is the vector with all zeros instead of a 1 in the ith position. The vectors given by expression 7.22 are called compressed coding vectors. So, the source packets injected into the network become

$$\begin{bmatrix} \mathbf{h}_i & |\mathbf{x}_i \end{bmatrix}^{\mathrm{T}} \tag{7.23}$$

and the packets transmitted by the intermediate nodes are

$$\mathbf{y} = \begin{bmatrix} \mathbf{y}_c & |\mathbf{y}_I \end{bmatrix}^{\mathrm{T}} \tag{7.24}$$

where $\mathbf{y}_c \in \mathbb{F}_q^p$ is the compressed coding vector and $\mathbf{y}_I$ is the payload, in the packet $\mathbf{y}$. Then, the relation between the coding vectors and their compressed version is

$$\mathbf{y}_c = \mathbf{y}\mathbf{H}^{\mathrm{T}}. \tag{7.25}$$

In particular, it is possible to recover the coding vectors from the compressed ones if the hamming weight $w(\mathbf{y}) \le r$.

On the receiver side, the sink has to recover the coding vectors of the packets from the compressed versions received to decode the system of linear equations. The fundamental idea is that if a code has minimum distance $d_{\min} = 2r + 1$, any set of $d_{\min} - 1 = 2r$ columns of $\mathbf{H}$ is linearly independent. The sufficient condition to make possible the decoding process is that the combination of each set of r columns gives a vector that is unique: therefore, the mapping 7.25 is an injective function. This permits to find the nonzero positions of the coding vectors $\mathbf{y}$. So, the knowledge of that and of the matrix $\mathbf{H}$ allows getting completely the linear coefficients of the classical coding vectors. There are two main approaches to find the nonzero positions in the coding vectors: the former is an exhaustive search among the all possible $\binom{h}{r}$ r-sets of nonzero positions, efficient for small h and r; the latter is to choose a code among BCH codes, Reed-Solomon codes, Goppa codes and algebraic geometric codes, for which is possible to use the Berlekamp-Massey algorithm. The main advantage of this approach is that the size of the coding vectors is reduced from h to $p = h - k$. The manner how the restriction in the number of combined packets influences the rate will depend on the topology of the wireless network, a positive aspect being that, normally, in wireless scenarios it is not suitable to allow linear combinations at all the nodes.

A further solution to the problem of compressing the coding vectors is to insert an array of g bits in the header called ID segment. At the ith source only the ith position is 1 and the others are 0s. At every intermediate node the ID segment of the outgoing packet is the XOR of the ID segment of the incoming packets. The ith element of the ith coding vector is not zero if and only if the ith bit in the ID segment is 1. The length of the ID segment is

$n/\log q$ symbols, so the total overhead becomes $(h - k + m)/\log q$. If the number r of sources combined together is fixed the overhead of the model above is $2r$ and the overhead of this ID segment scheme is $(m + n)/\log q$. Therefore, this scheme decreases the overhead for big values of r. A tutorial about network coding with compressed coding vectors can be found in [24].

7.6 Summary

Enhanced energy efficiency, reduction of complexity and reliability are all key attributes that network coding can deliver, and as such it is now considered an alternative approach to classical routing in wireless environments, particularly in very dynamically changing networks. This chapter intends to provide a pathway between the theory and the practical, since it is commonly viewed in the research community that the practical implementation of network coding is seen as the real bottleneck towards its adoption. To guide as along this path, we start with the fundamental overview on the theoretical framework of network coding for the design of efficient solutions in wireless scenarios. We specifically introduce hypergraph fundamentals and the main concepts of algebraic network coding adapted to this graph generalization. Then, we provide some key insights on the advantages and drawbacks of RLNC. Thereafter, we make the transition to real life applications and describe practical network coding issues, in which the three main paradigms are detailed: firstly, the solution which uses coding vectors and generations, simple but with the disadvantage of a significant overhead; secondly, a basic overview of noncoherent network coding and subspace coding theory, which represents the way to remove coding vectors; finally, the recent compression technique that keeps coding vectors and reduces their overhead. In particular, the last method has grown in importance after the discovery that the coding vectors bring passive information on network topology; which can be useful in real scenarios to find failures and bottlenecks.

7.7 Future Work

The exploration of network coding for wireless networks has still to face many open issues. In fact, now the well-known background of RLNC needs solutions to avoid the overhead given by the transmission of coding vectors. The discovery of subspace network codes and compressed coding vectors opened up new ways in the design of random linear network codes. At the moment, a key problem is the definition of the type of codes that are suitable for practical applications. Thus, it is necessary to substitute the initial concepts by coding solutions more efficient in terms of encoding and decoding complexity. Nevertheless, the reasons to keep coding vectors are also valuable and, therefore, research on further optimal methods to compress coding vectors is also very important. On the hand, energy efficiency and computational demand are important design attributes; the actual methods proposed demonstrated the potentials of network coding to be a strong candidate to replace routing, however practical implementation is still in its infancy.

Appendix

7.A Max-Flow Min-Cut Theorem

Let G be a graph that represents a single source unicast network. The source is transmitting information to the sink through the network. The aim is to maximize the flow between the source and the destination. The max-flow min-cut theorem is the result which demonstrates how to achieve this maximum possible flow. Before showing the statement of the theorem it is useful to provide some definitions.

The value of a flow is defined as the total flow into the sink. A cut of a s-t network is a set of edges such that if these edges are removed from the graph, the source node s and the sink node t are disconnected. Then, the capacity of a cut is the sum of the capacities of the edges in the cut.

Theorem 1 [3]. *The maximum possible flow from the source to the sink of the network is equal to the minimum value among all the cut-sets.*

Let us consider the example of the network in Figure 7.A.1. Among all the cut-sets, $\{d, e, f\}$, $\{b, c, e, g, h\}$ and $\{d, g, h, i\}$ are examples of them. The value of the cut-sets are for example 6 for $\{d, e, f\}$ and 10 for $\{b, c, e, g, h\}$. It is possible to verify that $\{d, e, f\}$ is the cut-set with minimum capacity and its capacity represents the maximum achievable flow between source and destination.

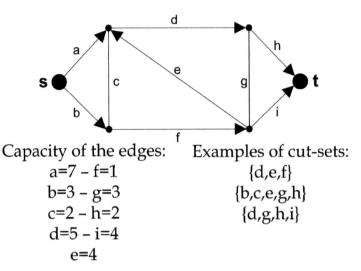

Capacity of the edges: Examples of cut-sets:
a=7 – f=1   {d,e,f}
b=3 – g=3   {b,c,e,g,h}
c=2 – h=2   {d,g,h,i}
d=5 – i=4
e=4

Figure 7.A.1 Representation of a network with a source and a sink. The values for the capacities are assigned to the edges and some examples of cut-sets are written below [3].

7.B Algebraic Preliminaries

This section will not be a formal presentation of algebraic concepts but only a brief description of some algebraic objects that were used above to show some network coding fundamentals.

A ring $(R, +, \cdot)$ is a set R closed under binary operations $+$ and $\cdot$ such that:

- $(R, +)$ is an Abelian group, which satisfies:
  - associative property: $a + (b + c) = (a + b) + c$ for all $a, b, c \in R$;
  - identity property: R has the identity element 0, that satisfies $0 + a = a + 0 = a$ for all $a \in R$;
  - inverse property: for each $a \in R$ there is an additive inverse element (the negative of a) such that $-a + a = a + (-a) = 0$;
  - commutative law: $a + b = b + a$ for all $a, b \in R$;
- The operation $\cdot$ is associative: $a \cdot (b \cdot c) = (a \cdot b) \cdot c$ for all $a, b, c \in R$;
- The distributive properties for multiplication over addition hold for all $a, b, c \in R$:
  - left distributive property: $a \cdot (b + c) = a \cdot b + a \cdot c$;
  - right distributive property: $(a + b) \cdot c = a \cdot c + b \cdot c$.

Then, a field $(K, +, \cdot)$ is a set K together with two binary operations $+$ and $\cdot$, such that:

- $(K, +, \cdot)$ is a ring;
- $(K - \{0\}, \cdot)$ is a commutative group.

Thus, a finite field $\mathbb{F}_q$ is a field with a finite number q of elements.

A fundamental object in algebraic geometry to study algebraic sets is the Grassmannian. A Grassmannian is a set $\mathcal{G}(k, n)$, with $0 \leq k \leq n$, which is the set of k-dimensional subspaces of the n-dimensional vector space $\mathcal{V}$ over the finite field $\mathbb{F}_q$. This object is especially used to analyze subspaces of a vector space without the necessity of defining any basis for $\mathcal{V}$. In particular, this concept becomes useful also for the definition of projective space. A projective space of order n over $\mathbb{F}_q$ is the set of all subspaces of the vector space $\mathcal{V} = \mathbb{F}_q^n$. Thus, a projective space is $\mathcal{P}_q(n) = \bigcup_{k=0}^{n} \mathcal{G}(k, n)$.

In the following, some structures that are fundamental in order to study the properties of subspaces are described. A matrix is in row echelon form if:

- All nonzero rows (rows with at least one nonzero element) are above any rows of all zeroes and all zero rows, if any, belong at the bottom of the matrix;
- The leading coefficient (the first nonzero number from the left, also called the pivot) of a nonzero row is always strictly to the right of the leading coefficient of the row above it;
- All entries in a column below a leading entry are zeroes, implied by the first two criteria.

So, a matrix is in reduced-row echelon form if it satisfies the further condition that every leading coefficient is 1 and is the only nonzero entry in its column. A Ferrers diagram depicts partitions as patterns of dots with the ith row having the same number of dots as the ith term in the partition. This kind of representation satisfies the following properties:

- The number of dots in a row is at most the number of dots in the previous row;
- All the dots are shifted to the right of the diagram.

Then, the number of row (or columns) of the Ferrers diagram is the number of dots in the rightmost column (top row) of the diagram. The echelon form of a vector $\mathbf{x}$ of length n and weight k, $EF(\mathbf{x})$, is the $k \times n$ matrix in reduced-row echelon form with leading entries of rows in the columns indexed by nonzero entries of $\mathbf{x}$ and dots in all entries which don't have terminals 0s or 1s. The dots of this matrix are the Ferrers diagram of $EF(\mathbf{x})$.

Then, the next example shows the last concepts above to make them clearer.

Example 1 [15]. Let consider the partition 6+5+5+3+2 of the integer 21. Hence, the 5×6 Ferrers diagram of this partition is

$$
\begin{matrix}
\bullet & \bullet & \bullet & \bullet & \bullet & \bullet \\
\bullet & \bullet & \bullet & \bullet & \bullet \\
\bullet & \bullet & \bullet & \bullet & \bullet \\
& \bullet & \bullet & \bullet \\
& & \bullet & \bullet
\end{matrix}
\tag{7.B.1}
$$

Now, let us consider the 7-dimensional vector $\mathbf{x} = (1, 0, 0, 1, 0, 0, 1)$. Its echelon form $EF(\mathbf{x})$ is the 3×7 matrix

$$
\begin{bmatrix}
1 & \bullet & \bullet & 0 & \bullet & \bullet & 0 \\
0 & 0 & 0 & 1 & \bullet & \bullet & 0 \\
0 & 0 & 0 & 0 & 0 & 0 & 1
\end{bmatrix}
\tag{7.B.2}
$$

which has 2×4 Ferrers diagram

$$
\begin{matrix}
\bullet & \bullet & \bullet & \bullet \\
& \bullet & \bullet
\end{matrix}
\tag{7.B.3}
$$

Acknowledgements

The research leading to these results has received funding from the European Community's Seventh Framework Programme (FP7/2007-2013) under grant agreements No. 264759 (GREENET), No. 285969 [CODELANCE] and Fundação para a Ciência e Tecnologia, through PTDC PTDC/EEA-TEL/119228/2010 – SMARTVISION. Victor Sucasas and Hugo Marques are PhD students at the University of Surrey.

References

[1] C. E. Shannon, A mathematical theory of communication, *SIGMOBILE Mob. Comput. Commun. Rev.*, vol. 5, no. 1, pp. 3–55, January 2001. [Online] available: http://doi.acm.org/10.1145/584091.584093.

[2] L. R. Ford and D. R. Fulkerson, Maximal flow through a network. *Canadian Journal of Mathematics*, vol. 8, pp. 399–404. [Online] Available: http://www.rand.org/pubs/papers/P605/.

[3] P. Elias, A. Feinstein, and C. E. Shannon, A note on the maximum flow through a network, *IRE Transactions on Information Theory*, vol. 2, no. 4, pp. 117–119, 1956. [Online] Available: http://dx.doi.org/10.1109/TIT.1956.1056816.

[4] R. Ahlswede, N. Cai, S.-Y. Li, and R. W. Yeung, Network information flow, *IEEE Transactions on Information Theory*, vol. 46, no. 4, pp. 1204–1216, July 2000.

[5] S.-Y. R. Li, R. W. Yeung, and N. Cai, Linear network coding, *IEEE Transactions on Information Theory*, vol. 49, no. 2, pp. 371–381, February 2003.

[6] R. Koetter and M. Médard, An algebraic approach to network coding, *IEEE/ACM Transactions on Networking*, vol. 11, no. 5, pp. 782–795, October 2003.

[7] T. Ho, M. Médard, R. Koetter, D. R. Karger, M. Effros, J. Shi, and B. Leong, A random linear network coding approach to multicast, *IEEE Transactions on Information Theory*, vol. 52, no. 10, pp. 4413–4430, October 2006.

[8] P. A. Chou, Y. Wu, and K. Jain, Practical network coding, 2003.

[9] R. Koetter and F. R. Kschischang, Coding for errors and erasures in random network coding, *IEEE Transactions on Information Theory*, vol. 54, no. 8, pp. 3579–3591, August 2008.

[10] D. Silva, F. R. Kschischang, and R. Koetter, A rank-metric approach to error control in random network coding, *IEEE Transactions on Information Theory*, vol. 54, no. 9, pp. 3951–3967, Septtember 2008.

[11] D. Silva and F. R. Kschischang, On metrics for error correction in network coding, *IEEE Transactions on Information Theory*, vol. 55, no. 12, pp. 5479–5490, December 2009.

[12] E. M. Gabidulin and M. Bossert, Codes for network coding, in *Proceedings of IEEE International Symposium on Information Theory, ISIT 2008*, pp. 867–870, July 2008.

[13] F. Manganiello, E. Gorla, and J. Rosenthal, Spread codes and spread decoding in network coding, in *Proceedings of IEEE International Symposium on Information Theory, ISIT 2008*, pp. 881–885, July 2008.

[14] V. Skachek, Recursive code construction for random networks, *IEEE Transactions on Information Theory*, vol. 56, no. 3, pp. 1378–1382, March 2010.

[15] T. Etzion and N. Silberstein, Error-correcting codes in projective spaces via rank-metric codes and ferrers diagrams, *IEEE Transactions on Information Theory*, vol. 55, no. 7, pp. 2909–2919, July 2009.

[16] A. Kohnert and S. Kurz, Construction of large constant dimension codes with a prescribed minimum distance, *CoRR*, vol. abs/0807.3212, 2008.

[17] M. Jafari, L. Keller, C. Fragouli, and K. Argyraki, Compressed network coding vectors, in *Proceedings of IEEE International Symposium on Information Theory, ISIT 2009*, pp. 109–113, July 2009.

[18] D. Lun, N. Ratnakar, R. Koetter, M. Médard, E. Ahmed, and H. Lee, Achieving minimum-cost multicast: A decentralized approach based on network coding, in *Proceedings of 24th Annual Joint Conference of the IEEE Computer and Communications Societies (INFOCOM2005)*, vol. 3, pp. 1607–1617, March 2005.

[19] D. Lun, M. Médard, and R. Koetter, Efficient operation of wireless packet networks using network coding, in *Proceedings of International Workshop on Convergent Technologies (IWCT)*, pp. 1–5, 2005.

[20] J. E. Wieselthier, G. D. Nguyen, and A. Ephremides, Energy-efficient broadcast and multicast trees in wireless networks, *Mobile Networks and Applications*, vol. 7, pp. 481–492, July 2002.

[21] Y. Wu, P. A. Chou, and S.-Y. Kung, Minimum-energy multicast in mobile ad hoc networks using network coding, *IEEE Transactions on Communications*, vol. 53, no. 11, pp. 1906–1918, November 2005.

[22] T. Ho and D. S. Lun, *Network Coding, An Introduction*. Cambridge University Press, 2008.

[23] A. Khaleghi, D. Silva, and F. R. Kschischang, Subspace codes, *Cryptography and Coding 2009*, pp. 1–21, 2009.

[24] C. Fragouli, Network coding: Beyond throughput benefits, *Proceedings of the IEEE*, vol. 99, no. 3, pp. 461–475, March 2011.

8

Secure and Energy Efficient Vertical Handovers

Hugo Marques[1,2], Joaquim Bastos[1,3], Jonathan Rodriguez[1],
and Rahim Tafazolli[2]

[1]*Instituto de Telecomunições, Aveiro, Portugal*
[2]*Universy of Surrey, Centre of Communication System Research, UK*
[3]*Universitat de Barcelona, Spain*
e-mail: vsucasas@av.it.pt

Abstract

The IEEE 802.21 standard, published in late 2008, offers generic link layer intelligence, independent of the underlying network access technologies, to facilitate handover optimization between heterogeneous access networks including IEEE 802 access technologies and other access technologies – a procedure known as vertical handover. Such handover can be triggered based on the maximization or minimization of specific parameters of the radio access, such as maximizing throughput or enhancing the energy efficiency of ongoing communications. Nonetheless vertical handovers raise security issues, even though there are security mechanisms for each access technology, a common security framework that enables authentication and key establishment from one technology to another is still missing. Latency associated with the security signalling during handover, especially the one related with network access authentication and authorization, is the most significant part of the entire handover latency between heterogeneous networks. Reducing this latency will assure service continuity during vertical handovers and enable the technology to reach its market potential. Towards this goal this chapter includes findings and achievements related to the procedures, protocols, research and, standardization activities – mainly focusing

Shahid Mumtaz and Jonathan Rodriguez (Eds.), Green Communication in 4G Wireless Systems, 139–176.

on work being developed by IEEE 802.21 [1] and IETF Handover Keying (HOKEY) [2] Working Groups. This chapter also includes a section with experimental results in an attempt to evaluate key aspects concerning energy efficiency in a VHO scenario modelled in a simulation platform based on the ns-2 network simulator, by simulating multiple VHOs under an IEEE 802.21 implementation. It is shown how the 802.21 functionality can be incorporated in ns-2 through add-on modules based in the 802.21 (draft 3) standard, developed by the National Institute of Standards and Technology (NIST).

Keywords: IEEE 802.21, MIH, handover, heterogenous networks, wireless networks, energy efficiency.

8.1 Introduction

Network access is emerging as an important issue in the current Internet world and a core concern for the Future Internet vision. Users often demand to be mobile and still be able to access their services any time, any place, and anywhere, independently of the networks and devices being used and the provider domains involved. Taking into consideration the plethora of available network access technologies, the transition between access networks (also known as handover procedure) has been identified as a critical point in system performance that should be properly designed. Handover between different access networks has been supported for some time in various technologies such as GSM, UMTS, 3GPP, 3GPP2 and Wi-Fi networks. Even though handovers could be executed between different access domains (e.g. roaming scenarios for cellular telephone operators and Wi-Fi networks [3]), these were only possible if made under the same access technology (known as horizontal handovers). In order to enable handovers between different access technologies (also known as vertical handovers), major standardisation efforts have taken place. When a VHO decision occurs, based on the input parameters set, a new radio access network (RAN) is chosen which targets to maximise or minimise specific parameters of the radio access, such as maximizing throughput or enhancing the energy efficiency of ongoing communications.

Several VHO decision algorithms were already studied and proposed by many authors over the years [4], contributing to the ultimate target of conveniently serving users, adopting the always best connected (ABC) concept [5]. These algorithms are designed mainly to ensure high user experience by adequately choosing the radio interface and access network for maintaining ongoing communications effectively, but can also address energy efficiency.

Anticipating that in the future vertical handovers would be the rule rather than the exception, an IEEE call for interest took place in the beginning of 2003 and in 2004 a Working Group (802.21) was formed. In the subsequent years the IEEE 802.21 WG had to solve the problem on how to seamlessly interconnect a wide variety of non-interoperable heterogeneous networks. Key obstacles to this task were [6]: (i) the lack of scalability due to limited support by operators; (ii) lack of standard handover interfaces; (iii) limited quality of service (QoS) guarantees during handover; and (iv) level of supported security when roaming across different access networks. Work within this group has driven to the release of its first standard on this topic, known as IEEE 802.21 Media Independent Handover Services [1]. Security was considered to be a hard task at that point in time [6] so it was not included in the IEEE 802.21 standard. Security extensions to Media Independent Handover Services are currently being dealt with by the IEEE 802.21 Security Task Group a (802.21a) [7]. The new IEEE 802.21a was published in May 2012 and is now freely available through the IEEE Get Program. Also contributing to IEEE 802.21a is the Handover Keying (hokey) Working Group [2]. More specifically HOKEY WG is providing assistance in the integration of EAP pre-authentication (also known as early authentication) within the IEEE 802.21a.

Parallel to this work, 3GPP defined the Evolved Packet Core (EPC) and a new component called Access Network Discovery and Selection Function (ANDSF) in the 3GPP standard [8], enabling it to interface with non-3GPP access network technologies. IEEE1900.4 has published IEEE 1900.4-2009 [9], that enables network-device distributed decision making for optimized radio resource usage in heterogeneous wireless access networks. IETF efforts towards mobility management produced several Mobile IP (MIP) solutions, such as Fast MIPv6 [10], Hierarchical MIPv6 [11] and Proxy MIPv6 [12] and also developed the Session Initiation Protocol (SIP) [13] that can be used as a mobility management solution.

8.2 Overview of IEEE 802.21

8.2.1 IEEE 802.21 Framework Functionalities and Communication Model

The IEEE 802.21 standard [1] offers generic link layer intelligence, independent of the underlying network access technologies, to facilitate handover optimization between heterogeneous access networks including IEEE 802

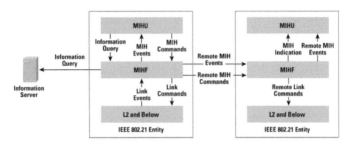

Figure 8.1   IEEE 802.21 MIHF Services [14].

access technologies and other access technologies. The "heart" of the standard is Media Independent Handover Function (MIHF), a new protocol layer located between Layer-2 (L2) and Layer-3 (L3) layers (often denoted as a L2.5 protocol layer). The intelligence of MIHF lies in the fact that it offers a complete set of command, event and information services to upper layer entities (MIH Users – MIHUs), as depicted in Figure 8.1. This allows for efficient control of the various link layers (Media Independent Command Service – MICS), detection of events corresponding to dynamic link changes (Media Independent Event Service – MIES) and information retrieval from network databases for sophisticated handover decisions (Media Independent Information Service – MIIS) respectively. For MIIS, a push model has also been specified [1].

These services are implemented through specific primitives that are grouped based on the Service Access Point (SAP) they use, see Figure 8.2. More specifically, communication between the MIHU and the MIHF is performed over the media independent SAP (MIH_SAP), while MIH_LINK_SAP handles MIH signalling exchange between the MIHF and the underlying link layers. Communication between peer MIH entities (often used by the MIIS) takes place over the MIH_NET_SAP. A small sample of SAP primitives is given in Figure 8.3.

In order to exchange MIH messages directly with a network MIH entity, the mobile should have connectivity with a point of attachment (PoA) at the network side and the respective network MIH entity should support point of service (PoS) functionality for the specific MN. The IEEE 802.21 standard defines a MIH PoS as a network-side MIHF instance that exchanges MIH messages with a MIHF located in a mobile device – a single MIH PoS can host more than one MIH service. As for a PoA it is defined as the network side endpoint of a layer 2 link that includes a mobile node as the other endpoint.

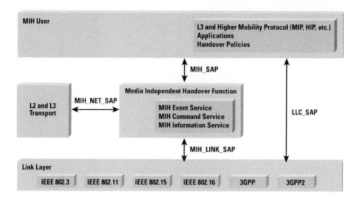

Figure 8.2 IEEE 802.21 MIHF Framework [14].

Figure 8.3 A sample of IEEE 802.21 MIH_LINK_SAP and MIH_SAP primitives [14].

The complete MIHF communication model can be seen in Figure 8.4, where different roles are assigned to the MIHF depending on its position in the system (at mobile device, serving point of attachment (PoA), candidate PoA, non PoA or, neither point of service (PoS) nor PoA). More specifically, the MIHF communications model defines the following reference points (RP) between these different instances of the MIHF [1]:

- RP1 and RP2 involve communication interfaces for layer 2 (L2), layer 3 (L3) and above;
- RP3 involves mainly communication interfaces for L3 and above. L2 interfaces are also possible for transport protocols such as Ethernet bridging and Multiprotocol Label Switching (MPLS);
- RP4 and RP5 involve only communication interfaces for L3 and above.

All these RPs serve as a conduit for the exchange MICS, MIES and MIIS related content.

8.2.2 IEEE 802.21 Vertical Handover Procedures

IEEE 802.21 standard defines the following vertical handover procedures:

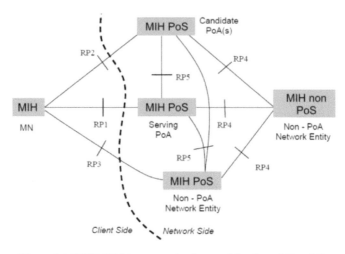

Figure 8.4 IEEE 802.21 communication model, adapted from [1].

- Handover initiation – Link unavailability;
- Handover preparation – Power on & scan;
- Handover preparation – Resource check;
- Handover preparation – Resource reservation;
- Handover preparation – L2 connection establishment;
- Handover execution – Complete.

In addition, the handover can be triggered either by the mobile (known as a mobile-initiated handover) or by the serving network (known as network-initiated handover). In a mobile-initiated handover (MIHO) a more detailed description of the above procedures could be as follows (the example assumes handover from a serving IEEE 802.16 (WiMAX) network to a target IEEE 802.11 (Wi-Fi) network):

1. *Measurement reports and handover initiation.* Initially, the mobile device configures and receives measurement reports about the WiMAX link. When the measured link parameters reach a pre-defined threshold, a *Link_Going_Down* event is received and the mobile device is alerted of an imminent handover;

2. *Handover preparation – information query to the MIIS server, network scan and resource availability check.* Upon deciding a handover is needed, the mobile device queries a MIIS server about possible candidate access networks (*MIH_Get_Information* message). This information request could be executed either periodically or only once; the IEEE

802.21 standard does not specify what the best approach is. After receiving the list of candidate networks, by the MIIS server, the mobile powers-on the Wi-Fi interface and performs scanning in order to find the listed Wi-Fi network. Moreover, it checks for available resources in the target network (Wi-Fi).

3. *Handover preparation – resource preparation.* After selecting Wi-Fi as the target network, the mobile asks from the WiMAX PoS to reserve resources at the Wi-Fi in order to prepare the handover to that network.

4. *Handover preparation – handover commitment.* In this phase the mobile executes the actual handover to the target network, causing a new layer 2 link establishment over Wi-Fi. At this point the mobile can either maintain the WiMAX interface active or shut it down.

5. *Handover completion – handover execution and resource release.* This is the last step and at this point the mobile needs to redirect all active data flows to the new interface (Wi-Fi). It first obtains network configuration (typical L3 configuration) and proceeds with mobile IP registration/binding (or other mobility management solution). After successful redirection, the mobile should release its resources in the old network. This can be achieved in two different ways, either by contacting the old network directly (needs the correspondent interface to be up), or by requesting the serving network (Wi-Fi PoS) to execute that step on its behalf.

8.2.3 Media Independent Information Services (MIIS)

MIIS are responsible for providing details on the characteristics and services made available by the serving and neighbouring networks. The use of MIIS gives mobiles a global view of available (all or filtered) networks in the area and allows a more effective link selection process which enables sophisticated handover decisions, for example, multi-objective connectivity decision optimisation. MIIS provide information by means of Information Elements (IEs) that can be logically grouped in IEs containers. Figure 8.5 shows some example IEs and containers (example is extracted from IEEE 802.21 standard [1]) a more detailed description of some of these IEs is provided in Table 8.1. As it can be seen in the depicted example, a mobile device may receive multiple candidate networks from multiple operators. Each operator can also operate multiple networks from different technologies. For each network type there is a list of PoAs and for each of these we can organize correspondent IEs in a PoA IE container. One key aspect

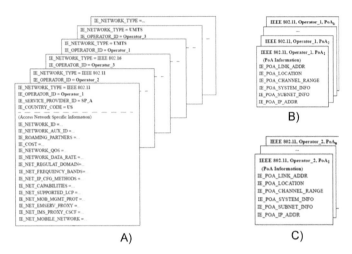

Figure 8.5 (a) MIIS list of candidate networks with general IEs; (b) and (c) PoA IE containers with information that describes a specific PoA (adapted from [1]).

Table 8.1 Some of the available IEs in MIIS.

IE_NETWORK_TYPE	Link types of the RANs available in a given geographical area
IE_COST	Indication of cost for service or network usage
IE_NETWORK_QOS	QoS characteristics of the link layer
IE_NETWORK_DATA_RATE	The maximum data rate value supported by RAN's link layer
IE_POA_LINK_ADDR	Link-layer address of PoA
IE_POA_LOCATION	Geographical location (e.g. coordinate-based, cell ID) of PoA
IE_POA_IP_ADDR	IP Address of PoA

of the MIIS, as currently implemented on IEEE 802.21, is that it provides mostly static information (cost, security and QoS capabilities, higher layer services supported, amongst others). Dynamic information (such as available resources) should be obtained only when that information is needed by the mobile and directly from the target network.

8.2.4 MIIS Discovery Mechanisms

MIIS information can be obtained through queries to the mobile's local database (based on information gathered from previous queries) and to remote Information Servers (known as MIIS servers or, in some literature, Mobility servers (MoS)), that may reside in a different network than the one currently in use. A mobile device can locate a MIIS server in two different ways:

1. The mobile has a pre-configured MIIS server address and does not need to run any type of discovery mechanism.
2. The mobile does not have a pre-configured MIIS server address and needs to use either DNS or DHCP discovery mechanisms. For this case the following scenarios apply:

   - Attempt to discover MIIS server in the home network – both DNS and DHCP discovery methods are applicable;
   - Attempt to discover MIIS server in the visited network – both DNS and DHCP discovery methods are applicable;
   - MN attempts to discover MIIS server in a 3rd party remote network – DNS must be used;
   - MN attempts to discover MIIS server in a roaming scenario – DNS should be used.

8.2.4.1 Using DHCP as a Discovering Mechanism

This approach has been studied by the Mobility for IP: Performance, Signalling and Handoff Optimization (MIPSHOP) Working Group [15] and is currently defined in RFC 5678 [16]. The MoS option (as it is commonly called by IETF) can be used to retrieve either the IP address or the domain name of the MIIS server. Figure 8.6 shows a possible message sequence chart for the discovery operation using DHCP. The server address can either be given as an IP address – "IPv4_Address-MoS" option – or as a fully qualified domain name (FQDN) – "IPv4_FQDN-Mo" option – the same applies for IPv6.

8.2.4.2 Using DNS as a Discovering Mechanism

The DNS discovery method [17] makes use of Naming Authority Pointer records (NAPTR) [18]. Since this records provide mappings to various kinds of information, the only ones relevant for DNS discovery are those with service fields "ID+M2X", where ID is the service identifier (IS, ES or CS) and X identifies a transport protocol supported by the domain (ex M2U for UDP, M2T for TCP and M2S for SCTP). As an example, an MN performs a DNS NAPTR query targeted to the surrey.ac.uk domain to discover the existence of a server providing IS services (queries are done using the service identifier "_IS" for the Information Service, "_ES" for the Event Service and "_CS"

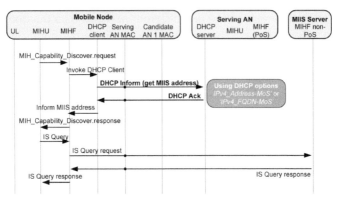

Figure 8.6 Locating IEEE 802.21 MIIS server using DHCP options.

Table 8.2 NAPTR record.

	order	pref	flags	service	regexp	replacement
IN NAPTR	50	50	"s"	"IS+M2T"	" "	_IS._tcp.su-rrey.ac.uk
IN NAPTR	90	0	"s"	"IS+M2U"	" "	_IS._udp.su-rrey.ac.uk

for Command Service). Assuming the DNS server returns the NAPTR record depicted in Table 8.2.

This indicates that the domain has a server providing IS services. This server supports both TCP and UDP, in that order of preference. If the client supports TCP and UDP, TCP will be used. However, at this point, the client does not know the FQDN or IP address of the desired server. In order to obtain this information, the client should send a DNS SRV query [19], targeted to _IS._tcp. surrey.ac.uk. A possible response from the DNS server is given in Table 8.3.

If the target is not an IP address (as in the above example) the MN needs to do an additional type A/AAAA DNS Query to resolve the name of the provided server (one at a time). Additionally, if the result of the SRV query contains a port number, then the MN should contact the server at that port number. If the SRV record did not contain a port number then the MN should

Table 8.3 DNS server response.

	Priority	Weight	Port	Target
IN SRV	0	1	XXXX	server1.surrey.ac.uk
IN SRV	0	2	XXXX	server1.surrey.ac.uk

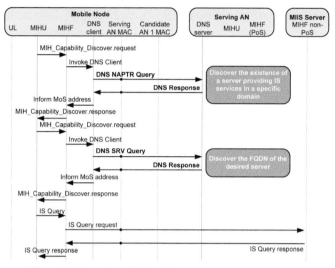

Figure 8.7 Locating IEEE 802.21 MIIS server using DNS.

contact the server at the default port number of that particular service. The default port number for MIH services, defined by IANA, is 4551 for both TCP and UDP [20]. Figure 8.7 shows a possible message sequence chart for the discovery operation using DNS.

8.3 Security Efforts towards IEEE 802.21 Vertical Handovers

The IEEE 802.21 standard provides extensible media access independent mechanisms that enable the optimization of handovers between heterogeneous IEEE 802 systems but does not address the security signalling required between such networks. According to Yoshihiro [21], security is crucial for IEEE 802.21 to reach its market potential. It is the vision of the IEEE 802.21 Security Study Group (TGa) [7] that a seamless mobility requires seamless security to make its applicability to government and enterprise networks. However, adding security related signalling typically adds considerable delays to handover efforts which lead to service interruption, affect real-time applications and, ultimately impact the user's experience. Handover scenarios contribute to this complexity, depending on whether the handover involves a change in the authentication server. With respect to this, the following scenarios can be considered:

- Handover between two different PoAs connected to same authentication server (same authentication realm). In this case, one of the following applies:
    - PoAs and the authentication server rely in the same network
    - PoAs act as proxy authenticators to the authentication server
- Handover between two PoAs connecting to two different authentication servers (different authentication realms).

For each of these scenarios, three different security concerns apply (with respect to IEEE 802.21):

- Security signalling optimization during handovers (includes mobile authentication and key establishment);
- Secure MIH message transport (including data integrity, replay protection, confidentiality and data origin authentication);
- Secure MIIS (MoS) server discovery.

8.3.1 Mobile Authentication and Key Establishment

Latency associated with the security signalling during handover, especially the one related with network access authentication and authorization, is the most significant part of the entire handover latency between heterogeneous networks. This is even more visible when the handover is performed between two different administrative domains that have neither trust agreement nor common infrastructure. It is therefore essential to reduce this latency to enable session continuity for applications. In the absence of an IEEE 802 standard dealing with this issue, IEEE 802.21 TGa has established a liaison with IETF's HOKEY WG towards the optimization for authentication and key establishment during VHOs, based on Extensible Authentication Protocol (EAP) [22] extensions – EAP is a generic framework supporting multiple authentication methods with the primary purpose being network access control. However, as stated by Lopez and Skarmeta [23], the handover latency introduced by full EAP authentication has proven to be higher than acceptable for real-time application scenarios. The current HOKEY Active Internet-Drafts and RFCs related to this issue are referenced in Table 8.4.

RFC 5169 [24] basically noted that the EAP framework was not suited for efficient handovers. The absence of re-authentication mechanisms mandated a full EAP method execution which caused unacceptable latency. Even though it was recognized there were attempts to solve this problem, those solutions failed because they were either EAP-method specific, EAP

Table 8.4 List of current HOKEY Active Internet-Drafts and RFCs related to fast & secure handovers.

Document	Title	Category	Status
RFC 5169	Handover Key Management and Re-Authentication Problem Statement	Informational	Informational
RFC 5295	Specification for the Derivation of Root Keys from an Extended Master Session Key (EMSK)	Standards track	Proposed Standard
RFC 5296 (possible update by draft-ietf-hokey-rfc5296bis-06)	EAP Extensions for EAP Re-authentication Protocol (ERP)	Standards track	Proposed Standard
RFC 5749	Distribution of EAP-Based Keys for Handover and Re-Authentication	Standards track	Proposed Standard
RFC 5836	Extensible Authentication Protocol (EAP) Early Authentication Problem Statement	Informational	Informational
draft-ietf-hokey-arch-design-11	Handover Keying (HOKEY) Architecture Design	Informational	RFC Ed Queue
draft-ietf-hokey-erp-aak-07	EAP Re-authentication Protocol Extensions for Authenticated Anticipatory Keying (ERP/AAK)	Standards track	Last call

lower-layer specific, not designed for scenarios involving handovers to new authenticators or they did not conform to the AAA (Authentication, Authorization, and Accounting) keying requirements specified in RFC4962 [25]. To cope with this, RFC 5169 defines the core design goals for a generic mechanism to reuse derived EAP keying material for handover, whose scope is specifically re-authentication and handover between authenticators within a single administrative domain.

RFC5295 [26] exploits the use of the Extended Master Session Key (EMSK) generated by an authenticated key exchange within the EAP framework to be used solely for deriving root keys. After a successful EAP authentication, both the peer device and the authentication server end up with two keys: a Master Session Key (MSK) and an EMSK. While the EAP framework clearly specifies the use for the MSK, it reserves the EMSK for future use. RFC 5295 uses the EMSK to derive keys for multiple use cases: Usage-Specific Root Key (USRK) and Domain-Specific Root Key (DSRK).

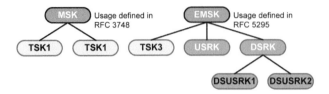

- **Master Session Key (MSK):** keying material derived between the EAP peer and server
- **Transient Session Key (TSK):** ephemerous keying material derived from the MSK (or EMSK) and used to protect data
- **Extended Master Session Key (EMSK):** additional keying material derived between the EAP peer and server
- **Usage-Specific Root Key (USRK):** keying material used for a particular usage definition
- **Domain-Specific Root Key (DSRK):** keying material restricted to use in a specific key management domain
- **Domain-Specific Usage-Specific Root Key (DSUSRK):** same as DSRK, but for a particular application

Figure 8.8   Keying hierarchy as defined in RFC 5295.

The root keys are cryptographically separate, meaning that given any root key it will be computationally infeasible to derive any of the other root keys or even the EMSK. This key-level hierarchy is depicted in Figure 8.8. Additional information to retain from this RFC is that root keys should only be used to derive child keys and never used to protect data, their lifetime should not be greater than that of the EMSK, and they should be named, as recommended by [25].

RFC2596 [27] addresses the problem statement presented in RFC 5196 by specifying EAP Re-authentication Extensions (ERXs) for efficient re-authentication using EAP. The EAP Re-authentication Protocol (ERP) is therefore the protocol that uses these extensions. ERP design includes a boot-strapping protocol, new EAP packets and updates the key hierarchy defined in RFC5295 as depicted in Figure 8.9. ERP is independent of lower layers and can re-authenticate a peer that has valid, unexpired key material from a previously performed EAP authentication, in a single round trip. ERP effectively reduces handover delay for intra-realm handovers by eliminating the need to run full EAP authentication with the home EAP server; however, in the case of inter-realm handover, ERP is not suitable, or an additional optimization mechanism is needed to transfer the keying material on the target PoA.

RFC5749 [28] proposes a solution for the distribution of EAP-based Keys for handover and re-authentication, since neither RFC5295 nor RFC5296 address this topic. Key distribution is an essential task in handovers, because neither the MSK nor the EMSK can leave the EAP server [26], root keys need to be sent (available) to other network servers. More specifically, RFC5479 defines a Key Delivering Server (KDS) and a Key Requesting Server (KRS),

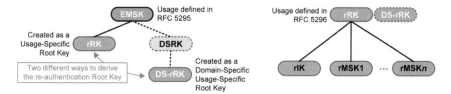

- **re-authentication Root Key (rRK):** keying material derived from the EMSK (or DSRK) and used to derive an rIK, and rMSKs
- **Domain-Specific rRK (DS-rRK):** same as rRK but restricted to a particular domain
- **re-authentication Integrity Key (rIK):** keying material derived from the rRK and used for integrity protecting the ERP exchange
- **re-authentication MSK (rMSK):** keying material derived derived at the peer and server following an ERP exchange

Figure 8.9 ERP keying hierarchy as defined in RFC 5296.

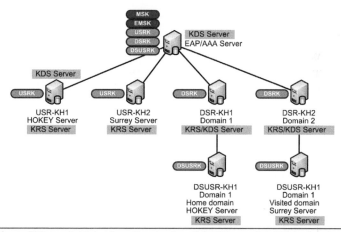

- **EAP/AAAServer:** EAP server that derives the root keys together with the AAA server that distributes these keys
- **USR-KHx:** network server that is authorized to request and receive a USRK from the EAP server
- **DSR-KHx:** network server that is authorized to request and receive a DSRK from the EAP server
- **DSUSR-KHx:** network server that is authorized to request and receive a DSUSRK from the EAP server
- **Key Delivering Server (KDS):** network server that holds an EMSK or DSRK and delivers root keys to a KRS
- **Key Requesting Server (KRS):** network server that shares an interface with a KDS and is authorized to request root keys

Figure 8.10 Key Delivery Architecture as defined in RFC 5749, adapted from [28].

as explained in Figure 8.10, that are part of a key distribution exchange (KDE) protocol for the root keys distribution used for re-authentication in ERP.

RFC 5836 [29] discusses the scenario where a mobile device establishes authenticated keying material on a (or multiple) target attachment point (TAP – similar to PoA in IEEE nomenclature) prior to its arrival. This scenario is known as EAP Early Authentication. The problem stated is focused on mitigating delay due to EAP authentication specifically in the handover execution (completion) phase between different link technologies and between different AAA realms. EAP early authentication mandates both authentication and authorization to be completed to a candidate attachment point (CAP – similar to candidate PoA) while the peer (mobile device) still has a valid connection to the serving attachment point (SAP-similar to serving PoA). This proactive behaviour is however limited to the scenarios where candidate authenticators can be easily discovered and accurate prediction of movement can be made. Focusing on the scenarios where early authentication is applicable while ERP does not work, the following usage models of early authentication are defined in RFC 5836:

- Direct pre-authentication: assumes the SAP is not involved in early authentication;
- Indirect pre-authentication: assumes the SAP interacts only with the CAP;
- Authenticated anticipatory keying (AAK) [30]: assumes the SAP interacts with the AAA server.

In any of these usage models, it is always assumed that there is no L2 connection between the peer and the CAP. Both direct and indirect pre-authentication need to execute a full EAP exchange before the handover takes place. To achieve this, direct pre-authentication assumes peers are capable of discovering CAPs, and route all communications through the SAP, whereas indirect pre-authentication makes use of an existing relationship between SAP and CAP, as it can be seen in Figure 8.11. Contrary to the two previous usage models, AAK does not execute a full EAP exchange and, like ERP, makes use of valid, unexpired keying material from a previously performed EAP authentication. AAK usage model is depicted in Figure 8.12 and, as it can be seen, CAP discovery is not the peer's responsibility. The CAP prediction problem (applicable to all usage models) could partially be solved by using comprehensive IEEE 802.21 IS (or similar approaches); however, EAP early authentication has one additional drawback – none of the currently existing AAA protocols distinguish between a normal authentication and an early authentication, which originates open issues. A complete list of such issues is included in [29, section 8].

A compilation of all the innovations/solutions proposed by the RFCs listed in Table 8.4 is attempted in the recent informational draft-ietf-hokey-arch-design-11 [32] HOKEY document. This candidate to RFC (currently in editorial queue) is the first attempt to design a handover key architecture (to be known as HOKEY architecture), that includes both re-authentication and early authentication approaches. The HOKEY architecture compiles four deployment scenarios addressed by the WG, extracted to Table 8.5, that include

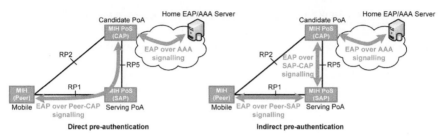

Figure 8.11 EAP early authentication direct vs. indirect pre-authentication models, adapted from [29–31].

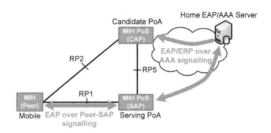

Figure 8.12 EAP early authentication authenticated anticipatory keying model.

the following components: Peer, Serving Authenticator, Candidate Authenticator, EAP Server and a EAP Re-authentication (ER) Server. The functions that must be supported within the architecture and the specific functionality of each of these components is thoroughly discussed in [32, sections 4, 5].

8.3.2 Secure MIH Message Transport

IEEE 802.21 base specification does not provide MIH protocol level security. The IEEE 802.21 Task Group a is currently developing work on MIH security issues to address this problem. Nevertheless we can assume MIH can typically run over TCP or UDP, so transactions can be protected by relying on the security of the underlying transport protocols (L2, L3) or through (to be developed) MIH specific mechanisms.

Protection at L2 could be achieved by following an approach similar to IEEE 802.11u [33], using the advanced security features provided by the Advanced Encryption Standard (AES). Protections at L2 however are most suitable for one hop links, if multiple L2 technologies exist between the two MIHF entities to be protected, than L2 security is not suitable.

Table 8.5 HOKEY Architecture usage scenarios (retrieved from [31]).

Simple Re-authentication	The peer remains stationary and re-authenticates to the original access point. Note that in this case, the SAP takes the role of the CAP.
Intra-domain Handover	The peer moves between two authenticators in the same domain. In this scenario, the peer communicates with the ER server via the ER authenticator within the same network.
Inter-domain handover	The peer moves between two different domains. In this scenario, the peer communicates with more than one ER server via one or two different ER authenticators. One ER server is located in the current network as the peer, one is located in the previous network from which the peer moves. Another ER server is located in the home network to which the peer belongs.
Inter-technology handover	The peer moves between two heterogeneous networks. In this scenario, the peer needs to support at least two access technologies. The coverage of two access technologies usually is overlapped during handover. In this case, only authentication corresponding to intra-domain handover is required, i.e., the peer can communicate with the same local ER server to complete authentication and obtain keying materials corresponding to the peer.

Protection at L3 can be achieved by using IPSec, which provides authentication, confidentiality, integrity, and anti-replay. IPsec support is mandatory for IPv6 and optional for IPv4. The IPSec Encapsulation Security Payload (ESP) can thus be used to authenticate MIH messages, provide confidentiality, integrity, and protection against replay attacks whereas Authentication Header (AH) can be used to provide authentication, integrity and anti-replay, but no confidentiality. IPSec is considered to be a complex protocol with too many roundtrips needed to establish a working Security Association (SA). A possible workaround is to use EAP keying material to derive new keys for IPSec, hence skipping first phase of IPSec protection (channel protection). Another limitation of the use of IPSec is related to dynamic IP addresses, which is certainty in vertical handover scenarios. The problem exists because IPSEC SAs are bound to static IP addresses, any change in the interface IP address will render a SA inoperable. In this case, the mobility and multihoming extension to Internet Key Exchange (IKEv2) protocol (MOBIKE) [34] may be used to establish SAs between two MIHFs. MOBIKE allows the IP addresses associated with IKEv2 and tunnel mode IPsec Security Associations to change.

Both L2 and L3 protection are not MIH specific, IPsec is however closer to MIH specific protection when compared to L2 protection for the broadest MIH scenarios.

Protection through MIH specific mechanisms will most certainly appear in the yet to be released IEEE 802.21a standard. It is expected that such mechanisms will be based on Transport Layer Security (TLS) [35] and Datagram TLS (DTLS) [36]. Further analysis on this topic is limited, since the recent draft versions (D04, D05 and D06) of IEEE 802.21a are not freely available for the general public.

8.3.3 Secure MIIS (MoS) Server Discovery

MIIS server discovery raises significant security issues. When using DHCP discovery method, it is important to remember that DHCP is inherently insecure, and it is recommended the use of DHCP authentication option as described in RFC 3118 [37]. If such option is not supported by the DHCP server, the protection will rely on the security provided by lower layers.

The use of DNS discovery method, like DHCP, is also inherently insecure. Crafting of DNS messages is possible, leading to man-in-the-middle and Denial-of-service attacks, resulting either in the inability to perform a handover, or perform the handover to an incorrect network. When using the DNS discovery method, it is recommended the use of DNS Security Extensions (DNSSEC) as described in RFC 4033 [38], and RFC 4641 [39] for best practices.

On another aspect, even if the MIIS server is securely discovered, there is still the question if it can be accessed before or only after the mobile (peer) is authenticated. Allowing unauthenticated access is a risk that opens a path to multiple attacks. Allowing only authenticated access creates a significant burden for the mobile in case multiple candidate PoAs are to be queried. A possible approach is to allow unauthenticated access only to obtain a subset of the IEs that describe a PoA (network), at least the ones needed for the handover policy function to determine if that PoA can be selected as a candidate PoA or not.

8.4 Simulation of IEEE 802.21 Handovers

This section describes an attempt to evaluate the reliability and scalability of the network simulator-2 (ns-2) [40] tool in simulating multiple vertical handovers under the scope of IEEE 802.21. Currently the 802.21 functionality can be incorporated in ns-2 by using external add-on modules developed by the National Institute of Standards and Technology (NIST) , that were based in IEEE 802.21 (draft 3).

8.4.1 Related Simulation Work

Since the release of the mobility package by NIST, as part of their joint work with IEEE 802.21 and the IETF, numerous research studies have used these modules. Rouil and Golmie [41] evaluates the performance of an adaptive channel scanning algorithm for IEEE 802.16 using a previous version of the WiMAX module that is used in this section. In [42] the handover latency for the cases where UDP and TCP carry MIH signalling messages is compared and some of the design trade-offs are presented. The evaluation of the performance of a vertical handoff scheme between 802.11 and 802.16 wireless access networks with respect to signalling cost, handoff delay and QoS support can be found in [43]. Mussabir [44] evaluates a proposed cross-layer mechanism for making intelligent handover decisions in FMIPv6 in terms of handover latency, packet loss and handover signalling overhead and Neves et al. [45] evaluate a new enhanced Media Independent Handover Framework (eMIHF) that allows for efficient provisioning and activation of QoS resources in the target radio access technology during the handover preparation phase.

The use of ns-2 with NIST modules has also been used to propose new implementation guidelines to the new security extension for IEEE 802.21 that is yet to come (IEEE 802.21a). Izquierdo et al. [46] compare different authentication techniques, namely, re-authentication and pre-authentication (early authentication), that may be used in order to reduce the time and resources required to perform a vertical handover. Izguierdo and Golmie [47] measure the performance of the authentication process in media independent handovers and considers the impact of using IEEE 802.21 link triggers to achieve seamless mobility and in [48] the authors propose an extension to the current network selection algorithms that takes into account security parameters and policies, and compares the handover performance with and without the proposed extension.

8.4.2 IEEE 802.21 Support in ns-2

Through the NIST add-on modules [49] it is currently possible to implement a limited version of the IEEE 802.21 handover specifications. These modules were developed for version 2.29 of ns-2. NIST added and changed numerous files in the standard release of ns-2 in order for it to support mobility scenarios. The following are some of these changes:

- Development of a new IEEE 802.21 add-on module [50], based on draft 3 of IEEE 802.21;
- Development of a new IEEE 802.16 add-on module [51], based on IEEE802.16g-2004 standard [52] and the mobility extension 802.16e-2005 [53] (both updated in 2007);
- Development of a new Neighbour Discovery add-on module [54] for IPv6 (limited functionality);
- Updated the existing IEEE 802.11 MAC implementation [55];

The IEEE 802.21 NIST add-on modules contain an implementation of the Media Independent Handover Function (MIHF) based on the draft 3 of IEEE 802.21 specification. The MIHF and MIH User are implemented as Agents. An Agent is a class defined in ns-2, extended by NIST, that allows communication with both the lower layers (i.e. MAC) and the higher layers (i.e. MIH Users), providing the mapping between the media independent interface service access point (MIH_SAP) and the media dependent interface (MIH_LINK_SAP and media specific primitives). Because of this, the MIHF can send layer 3 packets to remote MIHF and MIH User can register with the MIHF to receive events from local and remote interfaces. The MIHF is also responsible to get the list and status of local interfaces and control their behaviour. The MIH User class is hierarchy organized according to Figure 8.13. MIH Users make use of the functionalities provided by the MIHF in order to optimize the handover process. MIH Users typically will send commands to the MIHF and receive events or messages. The Interface Management class (IFMNGMT), depicted in Figure 8.13, provides flow management functions which facilitate the handover module in finding the flows that need to be redirected. The IFMNGMT also receives events from the Neighbor Discovery (ND) agent when a new prefix is detected or when it expires. The MIPV6Agent adds the redirection capability to the MIH User.

When a flow needs to be redirected, a message can be sent to the source node to inform him of the new address or interface to use. The handover class provides a template for handover modules and computes a new address after a successful handover.

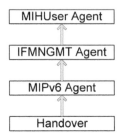

Figure 8.13 MIH User class hierarchy [50].

8.4.3 Supported Functionalities in ns-2 with NIST Add-on Modules

The current version of NIST add-on modules for ns-2 supports the following technologies in IEEE 802.21 scenarios: WiMAX (802.16), Wi-Fi (802.11), UMTS and ethernet (802.3). Integrating these technologies in the same node is not intuitive because external packages do not necessarily follow the same node structure as the one defined in the basic model (i.e. routing algorithms are different). To resolve this issue, NIST created the concept of *multiFace* node, which is a node who links to other nodes. The other nodes are considered interfaces for the *multiFace* node, and the *multiFace* node can be viewed as a "supernode'. This concept is illustrated in Figure 8.14. The interface nodes triggers events and send it to the *multiFace* node. The MIH Users on the *multiFace* node can register to receive these events. Additionally each of the interface nodes will also run an instance of the Neighbour Discovery (ND) agent in order to detect layer 3 movement. In order to identify wireless power boundaries to be used in the simulation, three variables were defined, see [55]. The description of the power boundaries is as follows:

- *CSTresh_*: defines the minimum power level to sense wireless packets and switch the MAC from idle to busy;
- *RX Tresh_*: defines the minimum power level to receive wireless packets without error;
- *pr_limit_*: is always equal or superior to 1 and is used in the following equation: *(RX Tresh_)*(pr_limit_)*; this equation defines the minimum power level that an interface senses before triggering a *Link_Going_Down* event. The higher the *pr_limit_* coefficient, the sooner the event will be generated.

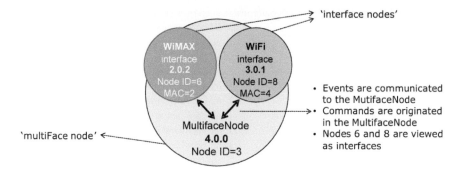

Figure 8.14 MultifaceNode – a node with multiple link layer interfaces.

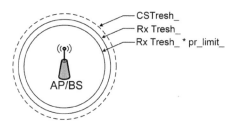

Figure 8.15 Wireless power boundaries in ns-2.

8.4.4 Support for Mobility and Integration of IEEE 802.21 IS, CS and ES

Support for subnet discovery and change of address when executing handover is possible through the use of *Router Advertisement* (RA), *Router Solicitation* (RS) and ND messages. RA messages are broadcasted periodically by Access Points (APs) or base stations (BSs) to inform MNs about the network prefix. In ns-2, each AP (Wi-Fi) or BS (WiMAX) is on a different subnet (domain or cluster in ns-2 nomenclature) and therefore will require a layer 3 handover, so the prefix is the address of the AP/BS who sends the RA.

The ND agent located in each MN is responsible for receiving RA messages and determine if it contains a new prefix; if yes, the ND agent informs the interface manager; if not, the timer associated with the current prefix is refreshed. If an MN loses the connection with its current PoA, AP or BS, it will not receive more RAs from that PoA and, therefore, the current prefix expires; in this case a notification is also sent to the interface manager. RS messages are used by MNs to discover new APs or BSs after the handover.

Table 8.6 HOKEY Architecture usage scenarios (retrieved from [31])

MIH commands	MIH commands
Link Event Subscribe	Link UP
Link Event Unsubscribe	Link Down
Link Configure Threshold	Link Going Down
Link Get Parameters	Link Detected
MIH Get Status	Link Event Rollback
MIH Link Scan	Link Parameters Report
	Link Handover Imminent
	Link Handover Complete

As for MIH services, IS are currently not supported in NIST add-on modules. Both CS an ES are supported, Table 8.6 shows the correspondent primitives.

8.4.5 Generalized Description of Handover Events in ns-2

To better understand the events generated in ns-2 when an MN executes a handover, the following experience was executed based on the following scenario description: A particular MN starts in a WiMAX cell and in the first 10 s the MN is stopped. After 10s it starts to move towards the centre of the WiMAX BS and, in its way it detects a Wi-Fi network. Since the used handover algorithm considers Wi-Fi a better network than WiMAX, the MN makes the handover to the Wi-Fi network approximately at $t = 22$ s. The MN stays in this network for approximately 7 s, when it senses it is losing coverage. Since the WiMAX signal is still available the MN executes a new HO to the WiMAX network (because it has no better network). The MN is the closest to the BS at instant $t = 35$ s, when the Rx power is at its highest value. The MN then moves away from the BS centre and stops at instant $t = 40$ s.

Running the simulation and parsing the simulation data allowed for the monitoring of the power received at the MN's interfaces, depicted in Figure 8.16, and the drawing of the message sequence chart for this specific scenario, depicted in Figure 8.17.

8.5 Experimental Approach in Energy Efficient VHOs

8.5.1 Interface energy models in ns-2

The ns-2 simulation tool natively supports energy modelling [56], to some extent, which has been enhanced over the released versions. The ns-2 energy

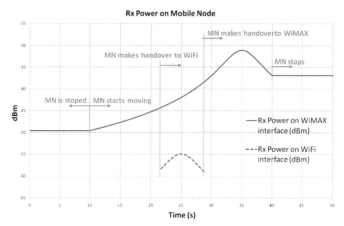

Figure 8.16 Power received at WiMAX and Wi-Fi interface in simulated scenario.

model is based on the *EnergyModel* class, which provides mechanisms to describe the energy consumption of an MN's wireless interfaces in the different states in which it can be. An MN is considered to have an energy budget (initially stored energy amount, e.g. in a battery), which will be consumed in the network operations it performs with each of its wireless interfaces. Each MN's interface is considered to be in one of the following states: *Transmit*, *Receive*, *Idle*, or *Sleep*. An interface can actually also be set to an *Off* state, but this implies a significantly longer transition time when the interface is requested to operate.

The *Transmit* and *Receive* states are self-explanatory. The *Idle* state describes an interface that is powered-on and ready to receive messages from the physical layer, while the *Sleep* state represents an energy saving mode in which the interface is unreachable, but with the immediate benefit of significant lower energy consumption. This model also considers the *Transition* state, which occurs when the interface transitions from *Sleep* state to *Idle* state. Each state is described by its own power consumption, which mainly depends on the physical layer being used. For instance, when the MAC layer performs a send operation, the interface initially in *Idle* state is transitioned to the *Transmit* state, ns-2 computes the time that is necessary to complete the operation, and multiplies that time by the pre-defined interface's power consumption in order to determine how much energy is used in the send operation. Afterwards, it subtracts from the MN's initial energy the estimated energy spent in the send operation, and so forth for every operation that occurs in the MN's interfaces. Moreover, when a *Transition* state occurs, the

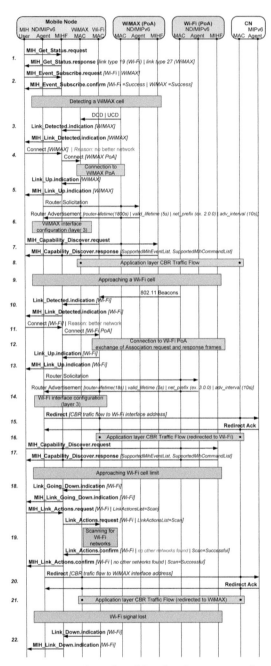

Figure 8.17 Message sequence chart describing handover events in an ns-2 simulated scenario.

Table 8.7 Power consumption for different interfaces

Power (mW)	WIMAX	Wi-Fi (802.11b)	UMTS
Transmit	532	890	1100
Receive	510	690	1100
Idle	80	256	555

transition power, considered here to be 200mW for any interface, is multiplied by the transition time, considered to be 5ms, to determine how much energy is spent while switching from *Sleep* to *Idle* state [57] -this is also subtracted from MN's energy budget. Table 8.7 presents current typical values for interfaces' power consumption [58–60].

8.5.2 Using MIIS for Energy Efficiency in ns-2

In an MN with multiple wireless interfaces, the fact that all interfaces are consuming power all the time represents a waste of energy which reflects in valuable battery lifetime loss, even if those interfaces are in idle mode since the MN is still powering the interface waiting to receive data. Hence it is possible to reduce energy wastage to a minimum, by selectively turning them off or putting them into a very low energy consumption state, e.g. sleep mode. This selective deactivation and activation of the MN's interfaces could be done according to context-aware information made available by the network, through MIIS. The required information is essentially the geo-location of the networks' BSs and APs. This information would allow the MN to decide, according to its own estimated position (e.g. from inbuilt GPS) and a selection algorithm, which interfaces should be active, or not, potentially allowing VHOs if justified. With this process a significant amount of energy would be saved, allowing extra mobility to MTs by extending its battery lifetime. In the results presented in Section 8.5.4, some primitives – specifically for the proposed energy saving approach – were implemented for requesting location information of BSs and APs, in ns-2 running IEEE 802.21 with NIST's add-on modules. A module to drive the state of the wireless interface was also developed and integrated to be able to switch the interface on and off based on context information regarding the location of the BSs, APs, and MNs.

Taking into consideration the described VHO between WiMAX and Wi-Fi in Section 8.4, an add-on targeting energy saving at the MN is depicted in Figure 8.18. Taking as reference Figure 8.17 after achieving *step 8*, once the MN is connected to the WiMAX network, and since no other network was detected and traffic flow is established, the MN can command its Wi-Fi interface to go into sleep mode. Once traffic flow is established between MN and the WiMAX network (*step 8*), upon MN's MIH User request, the MIIS client within the MN's MIHF communicates with the WiMAX BS's MIHF by sending a *MIH_Get_Information.Request*. The WiMAX BS's MIHF forwards this request to the MIHF where the MIIS server resides that, after generating a report based on the received query, returns the requested information through the appropriate *MIH_Get_Information.Response message*. This sequence of events is represented in Figure 8.18 as *step 8a*.

The implemented *query* requests the MIIS server for a list of available networks and respective PoAs in the surroundings of the WiMAX BS where the MN is connected, more specifically their link types (IE_NETWORK_TYPE), PoAs' link-layer address (IE_POA_LINK_ADDR), and PoAs' geographical (coordinate based) location (IE_POA_LOCATION). With that received information, the MN can determine according to its own location (e.g. provided by its inbuilt GPS chipset), speed and direction, when it should power down or wake up any of its interfaces. In the considered WiMAX"Wi-Fi"WiMAX VHO case (Figure 8.16) this happens approximately at $t = 20$ seconds when the MN determines that, according to its own position mapped onto the geo-referenced information provided by the MIIS on existing PoAs and the estimated power received, it should wake up its Wi-Fi interface. This happens just in time to proceed with *steps 9–18*. After traffic flow is successfully redirected to Wi-Fi network (*step 16*), the MN can put its WiMAX interface to sleep until further notice.

After receiving the Wi-Fi *Link Going Down* indication (*step 18*), and although the MN kept in its internal DB the list of nearby available networks and respective PoAs supplied by the MIIS server upon request, the Wi-Fi *Link Scan* procedure (*step 19*) is nevertheless commanded by the MN for searching other nearby Wi-Fi networks. This could prevent damage from a MIIS server incomplete database, and could also eventually allow MNs to help maintain the MIIS server DB up to date, by using additional appropriate information exchange. This was not considered to be in the scope of the present work, and therefore was not implemented. To note that the frequency of information exchange between MN and MIIS server should be enough to assure full knowledge of all the available networks in the MN's path, and it should depend mainly on MN's velocity.

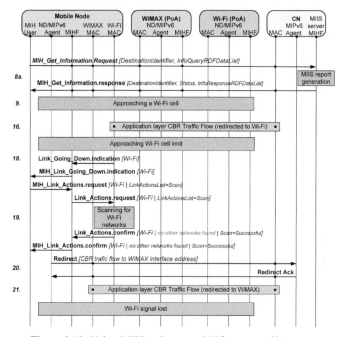

Figure 8.18 Using MIIS to improve VHO energy efficiency.

8.5.3 Simulated Scenario

The proposed approach for taking advantage of MIIS, targeting energy efficiency uses the MIIS primitives as described in the previous section, implemented over the MIHF functionality in ns-2. The simulated scenario and simulation parameters are depicted in Figure 8.19 and Table 8.8, respectively.

It consists essentially in a network topology composed by one WiMAX BS, 3 Wi-Fi APs and 25 MNs placed randomly, following a uniform distribution, inside the WiMAX cell. The WiMAX cell has a coverage radius of 500 meters, and each AP a coverage radius of 100 meters. Both WiMAX BS and Wi-Fi APs are wired to the core network through a router, and all wired links support 1 Gbps. All MNs have WiMAX and Wi-Fi interfaces, and also an inbuilt GPS module. The MNs initiate communication inside the WiMAX cell coverage area, where they remain static for a short period. After that, all MNs start moving with constant speed in a random, but steady, direction, which changes randomly after each simulated minute, following the random waypoint model [61]. All performed simulations considered a time period of 600 seconds, where MNs move with a specific constant speed.

Table 8.8 Power consumption for different interfaces.

Parameter	Value
WiMAX cell radius	500 m
WiMAX modulation	16QAM (10 Mbps)
WiMAX BS Tx power	15 W (42 dBm) @ 3.5 GHz
WiMAX RXThresh	1.215 nW (-59 dBm)
WiMAX CSThresh	80% of RXThresh
Wi-Fi hotspot radius	100m
Wi-Fi norm	IEEE 802.11b (11 Mbps)
Wi-Fi AP Tx power	100 mW (20 dBm) @ 2.417 GHz
Wi-Fi RXThresh	0,989 nW (-60 dBm)
Wi-Fi CSThresh	90% of RXThresh
Wi-Fi prlimit	1.2
Propagation channel model	Two-ray ground
Traffic bit rate	50 kbps (CBR)
Traffic type	UDP, full-duplex
MT's velocity	1, 2, 4, 8 m/s
MT's interfaces power consumption	(see Table 8.2)
GPS chipset power consumption	45 mW [62]

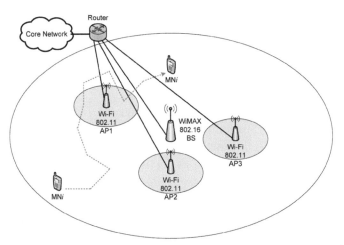

Figure 8.19 Simulated scenario for evaluating energy efficient VHOs by exploiting MIIS.

8.5.4 Simulation Results

As expected, since the GPS module consumes much less energy than any of the two radio interfaces, the total energy consumed during the considered simulation period is significantly lower than the traditional approach. Figure 8.20 (left) represents the total average energy consumed by each MN, taking into account both radio interfaces, using the proposed energy saving approach, or not, for different considered MNs' velocities. This figure includes the 95% confidence intervals superposed on the respective chart bars, which are minimal in the case when both interfaces are activated. It is obvious the consistent total energy consumed independently of the considered MN and its initial position, and path with intrinsic eventual VHOs. This happens since transmission power control was not implemented in the considered setup, and also because all MNs have both their radio interfaces turned on, even if in idle state, and are all transmitting and receiving at the same bit rate.

On the other hand, using the energy saving approach it is visible, namely through Figure 8.20 (right), the diverse energy consumption among the different MNs reflecting each one's path with intrinsic VHOs, and consequent deactivation of the alternative RAT interface, until it is foreseen to be needed. The reactivation of the Wi-Fi interface is done when an MN is less than twenty meters away from the coverage area of any AP, and the reactivation of a WiMAX interface is done when an MN is more than eighty meters away from any AP, before it is foreseen to be useful, according with the virtual map of PoAs detailed in each MN internal database, built on the exchanged information with the MIIS server.

Although it is not shown in this figure, it is clear that turning off the Wi-Fi interface is much more beneficial for energy saving, than powering off the WiMAX interface, since this interface consumes less energy in idle state. Once MNs move faster, the chances that they enter a different RAN coverage area are higher and thus the possibility of the occurrence of a HO. It is visible that the proposed approach is in average slightly more efficient for MNs moving faster, since they tend to be subject to more HOs, and therefore would in average benefit more from the proposed approach.

Figure 8.21 represents the average number of VHOs occurred involving an MN during the considered simulation period, together with the average amount of "smart" interface switching, for both Wi-Fi and WiMAX. This figure confirms what is expected in the considered scenario when MNs move faster, which on average effectively lead to a higher occurrence of VHOs. It is also visible that when using the proposed energy saving approach more interface switching happens at higher MNs' velocity, always targeting to optimise energy saving at the MNs' interfaces.

The average total time spent by the MNs on each RAN is represented in Figure 8.22, together with the average total accumulated time during which each of the radio interfaces was activated on the respective MN. As expected, the total time that a MN is connected to a Wi-Fi AP decreases with its

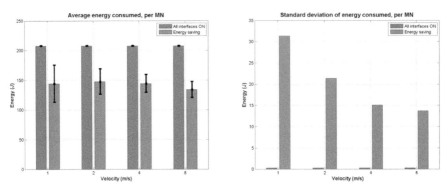

Figure 8.20 Average and std. deviation of energy consumed per MN, for different MNs' velocity.

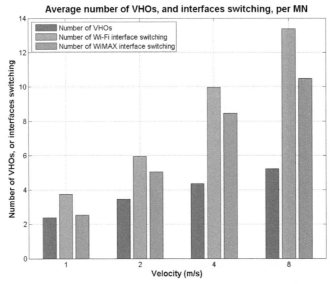

Figure 8.21 Average number of VHOs, and "smart" interfaces switching, for different MNs' velocity.

velocity, in the considered scenario setup, since the smaller coverage area can be crossed in a shorter time period. In this figure it is also visible the reduction in the accumulated time that the Wi-Fi interface is activated as the MN moves faster. This fact justifies the higher total energy saving for faster MNs, as previously shown in Figure 8.20, as the Wi-Fi interface consumes significantly more energy than the WiMAX interface, even in idle state.

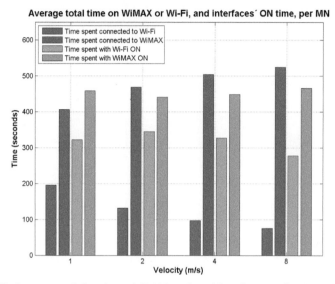

Figure 8.22 Average total time in each RAN, and total interfaces' active time, for different MNs.

From the carried out simulations, the total packet loss occurred in each simulation run, concerning each involved MN, was also accounted for, and the respective average and standard deviation results are presented in Figure 8.23 for the considered MNs' velocities. The average total number of packets transmitted and received in a simulation run by each MN is approximately 49,550. In its current implementation, the smart interface switching approach to energy saving sacrifices reliability of packet delivery in exchange for energy saving, since more packets are lost with this approach, namely increasing with MN's velocity. This packet loss was identified to occur namely at the reactivation of the sleeping interfaces, when the network association of the interface is delayed by channel congestion, in particular causing some packets to be lost when the WiMAX is not available and the Wi-Fi access point has bad channel conditions due to the distance to the MN.

Application-wise, this trade-off occurring with the proposed energy saving approach could eventually be tolerated in VoIP calls while the MNs are moving at lower speeds, but would lead to the degradation of video calls and streaming (low resolution MPEG4), which would become significantly damaged with the increasing of MNs' speed. The mitigation of this shortcoming should be possible by implementing a more intelligent interface switching

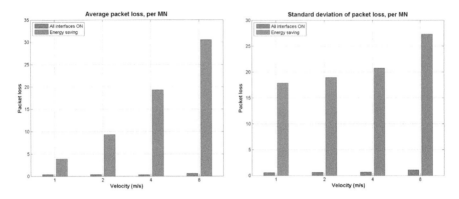

Figure 8.23 Average and std. deviation of packet loss, per MN, for different MNs' velocity.

algorithm also taking into account the speed of the MNs to enlarge the regions where both wireless interfaces are active.

8.6 Conclusions

It is expected that in the near future VHOs would be the rule rather the exception. Towards such a scenario the IEEE 802.21 standard has paved the way for manufacturers by specifying a set of media access-independent mechanisms that optimize handovers between heterogeneous IEEE 802 systems and between IEEE 802 systems and cellular systems. A key functionality of VHO must be its security as it would be crucial for IEEE 802.21 to reach its market potential. The procedures for a VHO can be split in multiple phases as explained in Section 8.2.2 and it is important to identify in which of such phases the security will be more crucial and also evaluate if the compromise of one handover phase will lead to the domino effect, resulting in the compromise of another phase. As explained in Section 8.3, major efforts have been made both by the IEEE (802.21a) and IETF (HOKEY WG), to deal with security in VHOs.

Another key functionality of VHO is the capability to address the recent concern for green technologies as it is boosting new research towards energy efficient systems. Some example energy saving techniques are proposed in this chapter by exploiting the MIIS. In vertical handover scenarios, the process of scanning for available candidate networks is time and energy consuming. The use of context awareness is very beneficial in such cases – to efficiently trigger the mobile's radio interfaces from idle mode and reduce the exhaustive search procedure for new channels. Current approaches for this problem consider that mobile devices should always save information gathered and network should always provide available context information to mobile devices. Under these guidelines, mobile devices can save even more energy by being aware of their surroundings, neighbouring mobile devices and access networks. The enabler for this approach would be a context-aware architecture that can push/pull rich context services to the mobile device. All these functionalities, however, demand more processing power and memory requirements in addition to the extra overhead caused by related data traffic. It is therefore important to draft a mechanism that will enable to measure if the cost of using a specific functionality will outweigh its benefits.

To simulate IEEE 802.21 VHOs, *ns-2* with NIST add-on modules proved to be a valuable tool not only to incorporate new modules (personalized for specific targets, as the energy saving simulations), but also to better understand the basic signalling of IEEE 802.21 standard. The use of NIST add-on modules however, only supports part of the standard (based on draft 3) as they were not conceived to simulate a high number of MNs; and some adaptations in *ns-2* were needed in order to run the simulations described here. Nevertheless, the obtained results have proven an acceptable approximation to what could be expected in real case scenarios.

The results attained through simulations using an *ns-2* based platform show that the proposed approach for energy saving can improve energy consumption, by approximately 30%, on average, with a fairly simple implementation. It is obvious that this energy saving can be even higher if a larger number of different radio interfaces are integrated in the MN. On the other hand, the currently proposed approach presents a trade-off in terms of additional packet loss, which is aggravated by MNs' velocity, and is mostly caused by delayed network re-association.

Acknowledgements

The research leading to these results has received funding from the European Community's Seventh Framework Programme (FP7/2007-2013) under grant agreement No. 285969 [CODELANCE] and Fundação para a Ciência e Tecnologia, through PTDC/EEA-TEL/119228/2010 – SMARTVISION. Hugo Marques and Joaquim Bastos are PhD students at the University of Surrey and Universitat de Barcelona, respectively.

References

[1] IEEE 802.21-2008, IEEE standard for local and metropolitan area networks – Media independent handover services, January 2009.

[2] IETF handover keying (Hokey) working group, http://datatracker.ietf.org/wg/hokey/charter/, accessed: 30/06/2012.

[3] IEEE.802-11F.2003, IEEE trial-use recommended practice for multi-vendor access point interoperability via an inter-access point protocol across distribution systems supporting IEEE 802.11 operation, 2003.

[4] X. Yan, Y. Ahmet Şekercioğlu, and S. Narayanan, A survey of vertical handover decision algorithms in fourth generation heterogeneous wireless networks, *Comput. Netw.*, vol. 54, no. 11, pp. 1848–1863, August 2010. [Online] Available: http://dx.doi.org/10.1016/j.comnet.2010.02.006.

[5] E. Gustafsson and A. Jonsson, Always best connected, *Wireless Communications, IEEE*, vol. 10, no. 1, pp. 49–55, February 2003.

[6] IEEE 802 tutorial: Handoff mechanisms and their role in IEEE 802 wireless standards, http://ieee802.org/16/tutorial/T80216-02_03.pdf/.

[7] IEEE 802.21 Task Group A (TGA), http://www.ieee802.org/21/TGa.html.

[8] Architecture enhancements for non-3GPP accesses, June 2010.

[9] IEEE P1900.4 Standard, Standard for architectural building blocks enabling network-device distributed decision making for optimized radio resource usage in heterogeneous wireless access networks.

[10] R. Koodli, Fast handovers for mobile IPV6, IETF, RFC 4068, July 2005.

[11] H. Soliman, C. Castelluccia, K. E. Malki, and L. Bellier, Hierarchical mobile IPV6 mobility management (HMIPV6), August 2005.

[12] S. Gundavelli (Ed.), *Proxy Mobile IPV6*, RFC 5213, IETF, August 2008.

[13] J. Rosenberg, H. Schulzrinne, G. Camarillo, A. Johnston, J. Peterson, R. Sparks, M. Handley, E. Schooler, Session initiation protocol, IETF, RFC 3261, June 2002.

[14] E. Piri and K. Pentikousis, IEEE 802.21, *The Internet Protocol Journal*, June 2009.

[15] Mobility for IP: Performance, signaling and handoff optimization (mipshop) working group, http://datatracker.ietf.org/wg/mipshop/charter/.

[16] G. Bajko and S. Das, Dynamic host configuration protocol (DHCPV4 and DHCPV6) options for IEEE 802.21 mobility server (MOS) discovery, RFC 5678, December 2009.

[17] G. Bajko, IEEE 802.21 mobility services using DNS, RFC 5679, December 2009.

[18] M. Mealling, Dynamic delegation discovery system (DDDS) part three: The domain name system (DNS) database, October 2002.

[19] A. Gulbrandsen, P. Vixie, and L. Esibov, A DNS RR for specifying the location of services (DNS SRV), RFC 2782, IETF, February 2000.

[20] T. Melia (Ed.), IEEE 802.21 mobility services framework design (MSFD), RFC 5677, December 2009.

[21] O. Yoshihiro, Five criteria for security extensions to media independent handover services, https://mentor.ieee.org/802.21/dcn/08/21-08-0007-00-0sec-802-21-security-5c.doc, January 2008.

[22] B. Aboba, L. Blunk, J. Vollbrecht, J. Carlson, and H. Levkowetz, Extensible authentication protocol (EAP), RFC 3748, June 2004.

[23] R. Lopez, A. Dutta, Y. Ohba, H. Schulzrinne, and A. Skarmeta, Network-layer assisted mechanism to optimize authentication delay during handoff in 802.11 networks, in *Proceedings of the 4th Annual International Conference on Mobile and Ubiquitous Systems: Computing, Networking and Services (MOBIQUITOUS 2007)*, 2007.

[24] T. Clancy, M. Nakhjiri, V. Narayanan, and L. Dondeti, Handover key management and re-authentication problem statement, RFC 5169, March 2008.

[25] R. Housley and B. Aboba, Guidance for authentication, authorization, and accounting (aaa) key management, July 2007.

[26] J. Salowey, L. Dondeti, V. Narayanan, and M. Nakhjiri, Specification for the derivation of root keys from an extended master session key (EMSK), RFC 5295, August 2008.

[27] V. Narayana and L. Dondeti, EAP extensions for eap re-authentication protocol (ERP), RFC 5296, August 2008.

[28] K. Hoeper, M. Nakhjiri, and Y. Ohba, Distribution of eap-based keys for handover and re-authentication, RFC 5479, March 2010.

[29] Y. Ohba, Q. Wu, and G. Zorn, Extensible authentication protocol (EAP) early authentication problem statement, RFC 5836, April 2010.

[30] Z. Cao, H. Deng, Q. Wu, and G. Zorn, EAP re-authentication protocol extensions for authenticated anticipatory keying (ERP/AAK), January 2012.

[31] S. Das, A. Dutta, and T. Kodama, Proactive authentication and MIH security, 5 July 2009.

[32] G. Zorn, Q. Wu, T. Taylor, Y. Nir, K. Hoeper, and S. Decugis, Handover keying (Hokey) architecture design, January 2012.

[33] IEEE P802.11u/D12.0, IEEE draft standard for information technology-telecommunications and information exchange between systems-local and metropolitan networks-specific requirements – Part II: Wireless lan medium access control (MAC) and physical layer (PHY) specifications: Amendment 7: Interworking with external networks, September 2010.

[34] P. Eronen, IKEv2 mobility and multihoming protocol (Mobike), RFC 4555, June 2006.

[35] T. Dierks and E. Rescorla, The transport layer security (TLS) protocol version 1.2, RFC 5246, August 2008.

[36] E. Rescorla and N. Modadugu, Datagram transport layer security, RFC 4347, April 2006.

[37] R. Droms and W. Arbaugh (Eds.), *Authentication for DHCP Messages*, RFC 3118, June 2001.

[38] R. Arends, R. Austein, M. Larson, D. Massey, and S. Rose, DNS security introduction and requirements, RFC 4033, March 2005.

[39] O. Kolkman and R. Gieben, DNSSEC operational practices, RFC 4641, September 2006.

[40] The network simulator tool, ns-2, http://nsnam.isi.edu/nsnam/.

[41] R. Rouil and N. Golmie, Adaptive channel scanning for IEEE 802.16e, in *Proceedings of Proceedings of 25th Annual Military Communications Conference (MILCOM 2006)*, Washington, D.C., October 23–25, 2006.

[42] D. Griffith, R. Rouil, and N. Golmie, Performance metrics for IEEE 802.21 media independent handover (MIH) signaling, *Wireless Personal Communications*, special issue on "Resource and Mobility Management and Cross-Layer Design for the Support of Multimedia Services in Heterogeneous Emerging Wireless Networks", 2008.

[43] Y. Zhang, W. Zhuang, and A. Saleh, Vertical handoff between 802.11 and 802.16 wireless access networks, in *Proceedings of IEEE Globecom'08*, November-December 2008.

[44] Q. Mussabir et al., Optimized FMIPv6 using IEEE 802.21 MIH services in vehicular networks, IEEE Transactions on Vehicular Technology, November 2007.

[45] P. Neves, F. Fontes, S. Sargento, M. Melo, and K. Pentikousis, Enhanced media independent handover framework, in *Proceedings of IEEE 69th Vehicular Technology Conference (VTC2009-z Spring)*, 2009.

[46] A. Izquierdo, N. golmie, K. Hoeper and L. Chen, Using the EAP framework for fast media independent handover.

[47] A. Izquierdo, N. Golmie, and R. Rouil, Optimizing authentication in media independent handovers using IEEE 802.21, in *Proceedings of 18th International Conference on Computer Communications and Networks*, 2009.

[48] A. Izquierdo and N. Golmie, Improving security information gathering with IEEE 802.21 to optimize handover performance, in *Proceedings of the 12th ACM international Conference on Modeling, Analysis and Simulation of Wireless and Mobile Systems*, Spain, October 2009.

[49] NIST ns-2 add-on modules for 802.21 (draft 3) support, `http://www.antd.nist.gov/seamlessandsecure/pubtool.shtml#tools`.

[50] The network simulator ns-2 NIST add-on – IEEE 802.21 model (based on IEEE p802.21/d03.00), January 2007.

[51] The network simulator ns-2 – NIST add-on – IEEE 802.16 model (MAC+PHY), January 2009.

[52] Ieee 802.16 wg, ieee standard for local and metropolitan area networks. part 16, Oct. 2004.

[53] IEEE 802.16 wg, Amendment to IEEE standard for local and metropolitan area networks. Part 16, December 2005.

[54] The network simulator ns-2 – NIST add-on – neighbor discovery, January 2007.

[55] The network simulator ns-2 – NIST add-on – mac 802.11, January 2007.

[56] The VINT project, the ns manual (formerly ns notes and documentation), `http://www.isi.edu/nsnam/ns/doc/ns_doc.pdf`, May 2010.

[57] V. Kakadia and W. Ye, Energy model update in ns-2, `http://www.isi.edu/ilense/software/smac/ns2_energy.html`, October 2011.

[58] F. Beaudoin et al., A fully integrated tri-band, MIMO transceiver RFIC for 802.16e, in *Proceedings of IEEE IEEE RFIC*, Atlanta, Georgia, USA, 15–17 June 2008.

[59] Y. Agarwal, R. Chandra, A. Wolman, P. Bahl, K. Chin, and R. Guptal, Wireless wakeups revisited: Energy management for VoIP over Wi-Fi smartphones, in *Proceeding of IEEE Mobisys07*, Puerto Rico, USA, pp. 179–191, June 2007.

[60] G. Lampropoulos, A.Kaloxylos, N. Passas, and L. Merakos, A power consumption analysis of tight-coupled WLAN/UMTS networks, in *Proceedings of 18th IEEE International Symposium PIMRC 2007*, pp. 1–5 ,2007.

[61] T. Camp, J. Boleng, and V. Davies, A survey of mobility models for ad hoc network research, *Wireless Communications and Mobile Computing*, vol. 2, pp. 483–502, 2002. DOI: 10.1002/wcm.72.

[62] Qualcomm Atheros Inc., AR1520 (GPS chip) specifications (AR1520-10-25-10), 2010.

9

Link Layer Modelling for Energy Efficient Performance Evaluation in Wireless Cellular Networks

Valdemar Monteiro[1,2], Shahid Mumtaz[1], Jonathan Rodriguez[1] and Christos Politis[2]

[1]*Instituto de Telecomunições, Aveiro, Portugal*
[2]*University of Kingston, Kingston upon Thames, UK*
e-mail: vmonteiro@av.it.pt

Abstract

Performance of cellular networks are typically classified in three key categories: link level, system level and network level, all of which require diverse experimental platforms and appropriate interfacing to ensure all the dynamics of the cellular environment are captured. In the next two sections we will provide the author with an overview towards the building blocks required to analyze and evaluate the system level performance of cellular systems, that can also have application towards generic wireless systems. In first instance, this chapter presents the link level building blocks and modeling approach, a vital ingredient that determines how the physical layer transmission scheme influences the quality of the transport channels, and consequently the system level ecosystem. The link level building block acts as the interface for the system level experimental platform that will be elaborated further in the sequel chapter.

Keywords: Link level, simulation, LTE, multi-RAT, beyond 3G.

Shahid Mumtaz and Jonathan Rodriguez (Eds.), Green Communication in 4G Wireless Systems, 177–201.

9.1 Introduction

The ever growing demand for high speed connectivity coupled with the fore-seen increase in data traffic have led wireless designers to rethink their design strategy for the new generation of mobile systems. The third and fourth generations are consequence of this demand. Along with data traffic demand, is the control on energy consumption as this is increasing along with this bandwidth and bitrates demand.

One of the most prominent aspects in research is the design requirements for energy efficient communication systems, since the energy bill is seen by most operators a major cost in the operational expenditure of future networks.

In wireless access, mainly two different layers are designed and evaluated – the link layer (PHY) and medium access control layer (MAC Layer), in separated ways, or in common, in an approach called cross-layers for algorithms optimization. The link layer deals with single ling aspects, which includes the whole transmission chain at the granularity of the bit level. It includes all modules responsible for the transmission and reception of the signal to either end of the link like modulation, multiplexing and channel coding. In this chapter we will analyse the modelling and performance evaluation in terms of link layer, both in terms of its general concept and in terms of energy consumption and optimization.

9.1.1 Modelling and Simulation of Wireless Networks

While some aspects in cellular networks may be evaluated analytically, test-beds or computer simulations are often the only way to proceed. However, developing the actual prototype is difficult and expensive without some a priori knowledge on expected performance. Moreover, measurement trials are very time consuming, and very difficult to explore a large parameter space. Therefore considering an intermediate step in terms of computer simulations is commonly applied method when it comes to evaluating the performance of cellular networks. The optimization of cellular networks can be done on many different levels and for a large parameter space. This includes static optimizations during the network planning process, as well as highly dynamic optimizations such as optimized queue management or scheduling decisions at the frame level. Many optimization problems that occur can be solved by means of simple problem specific heuristics. However, certain problems demand for a more generic solution approach.

Simulations in wireless and cellular networks can be classified in three categories, namely Link level, Network level, and System level, with respect to the targeted layers in the protocol stack and with respect to the considered level of abstraction. We will evaluate the Link level in this chapter and the system level in Chapter 10.

The choice of the right simulator category depends on the investigated problem and the desired results. Usually, it is advantageous to decouple simulators working at different levels of abstraction in order to reduce complexity. In this case, appropriate interfaces are required between the simulators. However, some problems may require the combination of different simulation levels in one simulator. Recently, network and system level simulations have grown in synergy, owing to the increasing complexity of MAC protocols and algorithms that increasingly act on , adjacent cells, or depend on adjacent cells in their performance. Examples include interference coordination algorithms or certain aspects of scheduling algorithms.

9.1.2 Design and Analysis of Energy Efficient Wireless Devices

Although much research has targeted the hardware power optimization, the new generation of wireless (3G, and 4G) rely significant part of its "performance efficiency" on the flexibility afforded in the network in terms of supporting mobility between heterogeneous systems, network sharing and coordinated multipoint transmission among other services. These services account for a high fraction of the power consumption due to the transport of data transmission over the wireless connections. Therefore, these facilities provide the necessity for an overlay technology and new energy metric that supports energy efficiency. The dynamic resource allocation functional block, part of the Medium Access Control (MAC) in the wireless network, offers a good opportunity to enforce significant energy savings, since here we can control the scheduling policy and transmission protocol, and properly identify which users should be served the network resources: since each user will impose a significant energy burden on the network.

In this chapter we present the analysis and simulation of link layer strategies for B3G/4G wireless systems, regarding as well reporting their energy consumption. More specifically the optimization in energy efficiency can be enforced in the link adaptation mechanism, by hybrid forward error correction (FEC) and with the Automated Repeat Request (ARQ) protocol. These key points are described hereafter along with their performance analysis at system. Finally, we also analyze alternative energy efficiency approaches for link layer protocols.

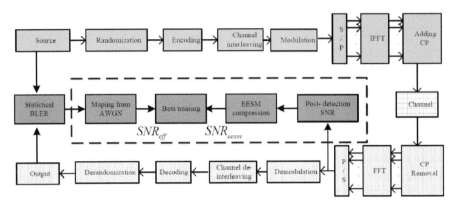

Figure 9.1 Overview of link level chain.

9.2 Energy Efficiency in the Link Layer

9.2.1 Link Layer architecture

Figure 9.1 shows the link layer chain of an OFDM based system, mainly composed by the coding and multiplexing steps. The chain includes randomization, channel coding, interleaving, modulation, FFT and Cyclic prefix. A WCDMA based link layer chain can be found in [1]. The figure also includes the interface model with the system level, shown by darker blocks. The link level chain is used to model the simulation of the chain, which computes BER/BLER rates in function of SINR in OFDM based systems. Effective SNR (SRNeff) and Exponential Effective SNR (SNReesm) used for SNR estimation in OFDM system modelling are described in detail in Section 9.3.3.

The output of the Link Level Simulator will be used as an input to the System Level Simulator, presented in Chapter 10. Link level simulators try to model the behaviour of the system in a computationally efficient way. Performing link level simulations has a high computational cost, therefore these simulations are normally performed in advance, and the results obtained are stored in a look-up table. Then, these results can be easily used to model the PHY behaviour together with higher layer protocol performance from a system perspective avoiding computational costs subsequent to online modelling of the physical layer in a system level simulator, for example. Table 9.1 summarizes the simulation parameter for Link level simulation and calibration results are shown in Figure 9.2.

Table 9.1 Link level simulator parameters.

Format	Beta (dB)
System Channel Bandwidth (MHz)	10
Sampling Frequency (F_p in MHz)	11.2
Subcarrier Frequency Spacing (f kHz)	10.94
FFT Size (N_{FFT})	1024
Coding	Convolutional (1/2,2/3,3/4,5/6)
	Turbo Coding(1/2,2/3,3/4,5/6)
	LDPC
Modulation Schemes	QPSK,16QAM, 64QAM
Channel	SISO/MIMO
	3GPP channel model (Ped B and VehA)
Null Subcarriers	184(DL), 184 (UL)
Pilot Subcarriers	120 (DL), 280 (UL)
Data Subcarriers	720 (DL), 576 (UL)
Data Subcarriers per Subchannel	24 (DL), 16(UL)
Number of Subchannels (N_S)	30 (DL), 35(UL)
Useful Symbol Time ($T_b = 1/f$) in μs	91.4
Guard Time ($T_g = T_b/8$) in μs	11.4
OFDM Symbol Duration ($T_s = T_b + T_g$) in μs	102.9
Number of OFDMA Symbols per frame (5 ms)	48
Data OFDM Symbols	44

9.2.2 Efficient Link Adaptation and Number of Re-transmissions

Link adaptation is used to adapt the transmission characteristics to the channel conditions. In early generations the most common link adaptation for the channel and interference characteristic was the power control. 3GPP [2] proposed a new approach in the link adaptation for the packet based HSPA to control the level of interference, where constant power transmission is stated and the link is adapted by the selection of suitable modulation and coding rates [3]. The channel quality is sensed and the block of data is transmitted according to the most suited transmission mode. The correct transmission mode will both maximize the amount of data transmission and at the same time minimize errors and subsequent re-transmissions. From an energy efficiency point-of-view, avoiding re-transmission mean direct energy saving.

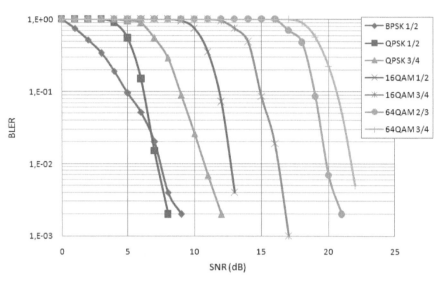

Figure 9.2 BLER vs. SNR plot for different modulation and coding profile on Veh A.

We present in the next sections two approaches that can be given for link adaptation for packet based communications.

9.2.2.1 Link Adaptation Algorithm 1 – BLER Threshold Based

In fact, a link adaptation algorithm can used for the selection of suitable Modulation and Coding Scheme (MCS) to be assumed in transmission of each packet. In this example, the selected MCS is the highest that corresponds to BLER values not higher than a threshold as described by the following expression:

$$MCS = \max \left(\arg_i \left(BLER(MCS_i) \leq BLER_{threshold} \right) \right) \qquad (9.1)$$

In case of services with variable packet sizes, the RRM algorithm could map packets of variable size into variable length radio blocks (see Section 10.3.4), which depends on the modulation and coding. BLER threshold is chosen so as to minimize both re-transmissions and energy consumption [4] (for instance, 3GPP proposed in UMTS a 10% BLER threshold).

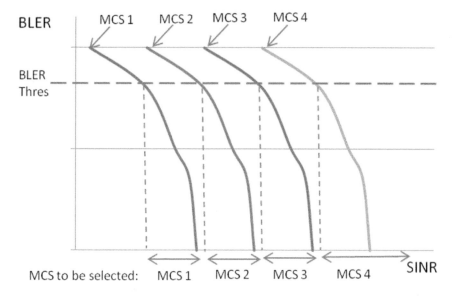

Figure 9.3 Modulation and coding selection, based in BLER threshold.

9.2.2.2 Link Adaptation Algorithm 2 – Throughput Maximization Based

For each MCS MCS_i, we define the predicted throughput $T_i : T_i = PBR(1 - BLER)$, where $PBR = Binfo/T_f$ is the Packet Bit Rate, $Binfo$ is the size of the frame/block in useful bits and T_f is the duration of the frame. The expression that translates this link adaptation algorithm is presented below, in equation (9.2). We can notice that this throughput formula only predicts accurately the throughput under the following assumptions:

- Constant channel and interference values during the frame period;
- Ideal ARQ (assuming no delay between various transmissions, and error free signalling);
- Simple ARQ (No combinations of transmissions).

We define the following parameters: SIR_i is the value of P_u/Ioc for which the T_i and $T_i + 1$ are equal. The following rules are applied to choose MS MCS:

- For $i = 1$ to 2, if $SIR_i \leq CI\_pred(nTf) \leq SIR_{i+1}$, $MCS_i + 1$ is selected for MS data transmission;
- If $CI\_pred(nTf) \leq SIR_1$ MCS1 is selected for MS data transmission;

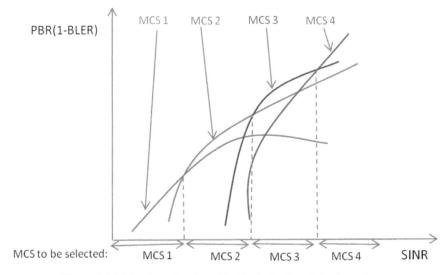

Figure 9.4 Link adaptation, based in Maximization of the throughput.

- If, MCS4 is selected for MS data transmission.

$$MCS = \max\left(\arg_i\left(Binfo(MCS_i) \times (1 - BLER_{Test})\right)\right) \qquad (9.2)$$

9.2.3 H-ARQ with Combining Methods

In this section, we present a partially asynchronous Hybrid Automated ReQuest mechanism. The ARQ mechanism is a stop-and-wait process responsible for frame error control and frame retransmission. The term Asynchronous Hybrid Automated ReQuest (H-ARQ) is used to describe the combination of ARQ and Forward Error Correction (FEC) to improve communication reliability. Common combining methods are Chase Combining [5], Incremental Redundancy [6] and Self Decodable Incremental Redundancy [5, 6], which improve communication reliability with multiple transmissions of block or in subsequent retransmissions of block information.

9.2.3.1 Chase Combining

The simplest form of Hybrid ARQ scheme was proposed by Chase [7], also Type III H-ARQ, with one redundancy version. The basic idea in Chase's combining scheme is to send a number of repeats of each coded data frame and allow the decoder to combine multiple received copies of the coded frame weighted by the SNR prior to decoding. This method provides diversity gain and is relatively simple to implement. The disadvantage is that the retransmission have retransmission has to be using the same modulation and coding rate, which may not be suitable for subsequent channel conditions.

H-ARQ is performed at the Radio Link Control (RLC) in the Radio Network Controller in the 3GPP specifications for the 3G and LTE systems. In this section, we present an implementation of the In this section, we present an ARQ mechanism with maximum number of re-transmissions N_{rmax} equal to 2. However, the presented case can be easily extended to larger number of re-transmissions. We recall that one data block is mapped onto one RU which is defined by its coordinates (one specified time slot and one specified spreading code).

In order to describe the protocol, we consider a particular Mobile Station MS_i having a connection with a particular Base Station BS_j. BS_j sends data to MS_i using channel HARQ SCH number 'n'.

The following flowchart is applied to describe an implementation of Chase combining:

1. 1st transmission

   - Using Dynamic Resource Allocation mechanism, BS_j schedules to send a set of data blocks to MS_i on H-ARQ SCH 'n', on a specified set of RU(s), with one specified MCS.
   - BS_j broadcasts signalling: BS_j warns MS_i that a first transmission is scheduled for the next sub-frame (which is dedicated to H-ARQ SCH 'n'). BS_j also indicates which RU(s) and MCS are dedicated to MS_i.
   - BS_j transmits data on the sub-frame dedicated to H-ARQ SCH 'n', keeping a copy in BS_j buffers dedicated to H-ARQ SCH 'n'.
   - MSi receives data, processes it and then extracts erroneous data blocks (identified thanks to CRC check). If there are no erroneous data blocks, then MS_i sends an "ACK" message to BS_j. If there are erroneous data blocks, then MS_i sends a "NACK" message with RU(s) coordinates of erroneous data blocks, MS_i stores erroneous data blocks in buffers dedicated to HARQ SCH 'n', and waits for their retransmission during the next sub-frame dedicated to H-ARQ SCH 'n'. Each stored data block is the "1st version" the original data block.

2. Retransmission

- This phase occurs after BS_j has received a NACK message. Thanks to the NACK message, BS_j has identified the subset of data blocks to be re-transmitted.
- BS_j empties all buffers dedicated to H-ARQ SCH 'n' except buffers containing data blocks to be re-transmitted. BS_j schedules the subset of data blocks. To each data block of the subset, BS_j allocates the same Ru and the same MCS as for the 1st transmission.
- No need to broadcast signalling because MS_i knows that a retransmission is planned and which RU(s) are concerned.
- BSi re-transmits data.
- MS_i receives the re-transmitted data, and combines each received data block with its "1st version" (Chase Combining). The result of this combination is processed. Then, MS extracts erroneous combined data blocks (identified thanks to CRC check). If there are no erroneous combined data blocks then, MS_i sends an ACK message to BS_j. If there are erroneous combined data blocks, then MS_i stores them in buffers dedicated to HARQ SCH 'n' and sends a NACK message with RU(s) coordinates of erroneous data blocks. Each stored data block is the "2nd version" the original data block.

9.2.3.2 Incremental Redundancy

Incremental redundancy (IR) is an H-ARQ technique, also called ARQ type II, where additional redundant information is incrementally transmitted if the decoding fails on the first attempt. This method, is more complex to implement in the receiver when compared to Chase combining, but it has the advantage that the new transmission can be adapted to actual and channel conditions of the new transmission. In this method, the achieving coding gain is obtained by adding redundant information in the retransmitted data block instead of repeating same block. Prior to the decoding, the receiver combines the subsequent received blocks with the first received. As disadvantage this method cannot decode the retransmission data separately.

9.2.3.3 Self Decodable IR

An approach to the Incremental Redundancy method called Self decodable IR has been proposed to face the disadvantage of the IR. In the Self decoded IR method, redundant information is added to the first data block and single versions are produced in the retransmissions applying some puncture patterns. If the channel is in good conditions in the retransmission period, the receiver is able to decode all versions separately. If no particular conditions of quality is verified, the transmitted, versions are combined according to their puncturing patterns and then decoded.

9.3 Link Layer Simulation and Interface to Upper Layer

9.3.1 Link Level Interface Modelling – General Concepts

The performance evaluation of a practical system by means of simulations has to consider the overall system layers of the whole communication protocol stack: physical layer, link layer (both Logical Link Control – LLC and Medium Access Control – MAC) and upper layers (network, application and transport layers). Simulations carried out on the radio link level are performed for a point-to-point link between the base and mobile stations, either in a SISO or MIMO propagation channel. The whole transmission chain is simulated at the granularity of the bit level. It includes all modules responsible for the transmission and reception of the signal to either end of the link. The ultimate goal is the generation of a set of curves illustrating the variation of the Bit Error Rate (BER), Block Error Rate (BLER) or Symbol Error Rate (SER) with the Signal to Noise Ratio at the bit level: E_b/N_0.

At the system level, simulations are conducted for a group of base stations, in a typical hexagonal cellular layout, which transmit (downlink connection) and receive (uplink connection) to/from a group of mobile stations attached to its area of coverage (cell). At this level simulations are conducted in a point-to-multi-point configuration, where a group of mobile stations are attached to each cell in the network and the ultimate goal is the generation of a set of metrics, which reflects the performance of the network in terms of: achieved user and cell throughput, packet drop ratio, average packet delay, etc.

Both layers perform simulations under different time-scales: physical layer simulations are performed at the bit level and system level simulations are performed at the frame interval, or transmission time interval. Although a hypothetical and desirable scenario in terms of the accuracy and validation of the results obtained, it is not practical in terms of complexity and simulation time to simulate the whole physical link between the base station and a single mobile station for all mobile stations in the network. The integration of all layers functionalities would be a huge task for the simulator and it would definitely take a huge amount of time, especially if several scenarios such as: traffic type, environment and cellular layout have to be considered. Therefore, there is a need to adopt a simple model that would be accurate enough to capture the signal statistics and impact on performance metrics, whilst still maintaining the simulation time frame within an acceptable limit.

For this reason, system performance evaluation is based on a separation among the different layers functionalities:

- On one hand system level performance evaluations do not consider the steps performed on the physical layer for the transmission of each bit of information between both ends of the communication link. System Link level Radio Resource Management (RRM) algorithms performing at the frame interval level of granularity consider the physical layer as a "black box", interacting with it by means of well-defined interfaces.
- On the other hand, physical layer simulations are unaware of the algorithms performed on higher layers for RRM, such as Dynamic Channel Allocation (DCA), Power Control, Handover, Connection Admission Control (CAC) and Scheduling.

This separation among layers implies the definition of interfaces which must be properly designed in order to, as accurately as possible, affect system simulation results with the performance of the physical layer. This strategy results in an implicit trade-off: on one hand it allows obtaining results in an acceptable time interval and, on the other, there is the drawback of loosing accuracy in the obtained results and accuracy of the physical layer performance as seen by upper layers.

The performance of the physical layer is modelled by means of Look-Up Tables (LUT) in which the behaviour of the radio link is encapsulated. An example of a metric that could be used in the performance abstraction of the physical layer is the variation of the Frame Error Rate (FER) with the Signal-to-Interference plus Noise Ratio (SINR), averaged over many channel realizations for the specific channel model used. According to the simulated scenario two types of interfaces can be defined in physical layer abstraction for system level simulations [8, 9]:

Average Value Interface – this type of interface reflects the radio link quality for a long time interval. This scenario is typical for mobile speeds corresponding to values of the coherence time smaller than the duration of a single transmission time interval, making it unrealistically to assume the channel constant along one or two radio frames. Only statistical channel behaviour is assumed as channel state value is averaged over time. The average value interface is not accurate if there are fast changes in the interference due to, e.g., high bit rate packet users.

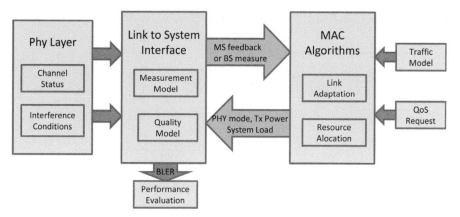

Figure 9.5 Overview of link to system level interface.

Actual Value Interface – this type of interface reflects the instantaneous value of the radio link. It is suitable for scenarios of low mobility, resulting in a slow fading channel profile.

Figure 9.5 illustrates the processing blocks involved in system level simulations and their dependencies.

These blocks reflect the functionalities implemented in both PHY (left) and MAC (right) layers. It can be observed that the performance of the physical layer, as modelled by the Link to System Interface Module (LSIM), depends on both the channel quality and the interference conditions, and is reflected in the computation of the SINR metric. The SINR depends on the position of each mobile on the network (geometric factor) and on the traffic load due to the total amount of active mobiles in the system. On the System Level the MAC sub-layer schedules the amount of resources required by each service flow. Its performance depends on the Quality and Measurement Models used. The Quality model is responsible for the estimation of the link performance based on the resource allocation. Link quality estimation is performed by means of the Packet Error Rate (PER) or the Block Error Rate (BLER) and these metrics are the outputs of the Look-Up Tables (LUT). The Measurement Model is the block responsible for the computation of the quality metric used by MAC algorithms such as: channel-dependent scheduling and resource allocation, power control and link adaptation.

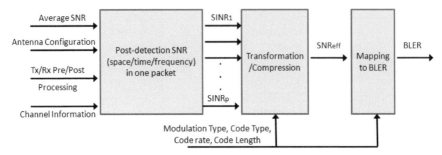

Figure 9.6 PHY-abstraction model.

9.3.2 Link Layer Abstraction Procedure

The abstraction method adopted in this chapter is based on the Exponential effective SNR mapping (EESM) algorithm that is based on actual value interface measurements, and especially suited for OFDM systems. The abstraction model (EESM) introduced is based on the concept of effective SNR. It takes the instantaneous channel and interference characteristics and other configurations into account and can provide the instantaneous performance of the real PHY layer system more accurately. The model consists of three main blocks [10] (see Figure 9.6).

- Average SNR → Post-detection SNR.
- Post-detection SNR → Effective SNR.
- Effective SNR → BLER.

The first block covers the calculation of quality measures $SINR_p$ for all relevant Resource Elements. (p is the total number of resource).To facilitate further processing an appropriate transformation and compression of the $SINR_p$ to one single parameter SNR_{eff} is performed within the second block. The SNR_{eff} (or an equivalent metric BLER code word) is finally mapped to the Block Error Rate (BLER). The goal is to cover the resource and power-allocation, channel impacts and pre-/postprocessing as well the first component i.e. by the extraction of the $SINR_p$. The following two steps have to account for the modulation and coding parameters. They should be designed in such a way that the final BLER mapping effectively models the performance of coding. In the following, the realization of the three parts will be explained in detail.

Block 1: Average SNR → Post-detection SNR
Basically the instantaneous SNR is chosen as a quality measure at System

Level. In the first step, the computation has to cover at least a sufficient number of samples with respect to channel coherence conditions for the user of interest and its interferers as well. Therefore, the interface is capable to account for the instantaneous channel conditions (including fast fading). The SNR has to be understood as a post receiver Signal to Noise and Interference Ratio, i.e. after detection etc. In the following equation, sig P stands for the power of the expected signals and noise P stands for the power of white Gaussian noise involved in the system

$$SINR_p = \frac{P_{sig}}{P_{noise}} \tag{9.3}$$

Post-detection SINR is calculated for each resource element, which means one for each time point and each frequency point. One packet is assumed to consist of p constellation points in the time/frequency space. All the SNRp values belonging to a particular packet are transformed/compressed/averaged to a single effective SNR. For a given channel allocation scheme one can reduce the number of SNRp BLER code word according to the channel conditions, i.e. with respect to the variation of the fading in the time and frequency domain (coherence time and coherence bandwidth for frequency selective fading).

Block 2: Post-detection SNR → Effective SNR
The second step of this modelling approach is considered the most essential part. In this step, all the $SINR_p$ values belonging to a particular packet are transformed/compressed/averaged to a single effective SINR. One can say that it is another kind of averaging which is different from linear calculating. For a given convolutional code and packet length N, the relationship between the packet error probability and signal-to-noise ratio can be determined for a conventional AWGN channel (with a constant SNR, via analysis or simulation. We assume that this relationship is known and represented as

$$PER_{AWGN} = h(SNR) \tag{9.4}$$

For a given L-dimensional vector SNR, which takes the instantaneous channel and interference characteristics and other configurations into account, we define the scalar effective SNR, as the SNR in an equivalent, constant signal-to-noise ratio AWGN channel which would yield the same packet error probability [11]. Thus, we relabel the relationship as:

$$PER = f(SINR) = h(SNR_{eff})$$

$$SNR_{eff} = g(SINR) \tag{9.5}$$

The calculation of effective SNR is the most essential part, and it should cover issues such as current channel condition, configuration of the antennas and the pre-and post- data processing in the real system so that the later processing of the model is independent of the channel. The effective SNR is brought out as an ideal concept, and there have been many ways to approximate the g function to approximate this concept, and EESM is one of them, which is explained in detail later in this chapter.

Block 3: Effective SNR → BLER
As mentioned above, the effective SNR is defined as the SNR in an equi-valent, constant signal-to-noise ratio AWGN channel which would yield the same packet error probability. It is an ideal concept and the calculation ap-proach described above is considered a good approximation for that. The mapping from mutual information to SNR, ideal and approximation should be simulated in the simple AWGN channel.

Since some adjustments have been done accordingly for every kind of modulation type in the second step explained above, the processing in the third step is expected to be independent of modulation scheme. That iss to say, the mapping curves should be generated BLER – Code Triple (Coding Type, Code Rate and Code Word Length), without the consideration of modulation scheme. And also the results are kept with high resolution and can be sub-sampled according to the accuracy needed in the simulation.

Following the three steps presented above, we can get the final block error rate for current configuration, channel condition, modulation and code scheme and so on. When HARQ with CC is introduced in the link layer, the post-detection SNR in the first step shall be calculated separately and combined first to get a new value as the input for the abstraction process. The other parts of the abstraction procedure remain the same.

9.3.3 SIR Estimation on OFDM Systems

In this section we will analyse into details the EESM method, the Link layer abstraction procedure on OFDM based systems for interface to system level modeling.

9.3.3.1 Effective SIR Mapping Functions

The role of an SINR-based mapping function is to provide the expected BLER/FER of a coded block of information as a function of the vector of SINR values of its data symbols, which can be transmitted through different types of resources (time, frequency, codes or spatial beams) depending on the type of multiple access implemented in the air interface. This calculation may involve the computation of the weights of both pre and post processing matrices, at the transmitter and the receiver, as for example whenever beamforming or spatial multiplexing is used in connection with MIMO.

The vector of instantaneous SINR values, associated to the resources assigned to the transmission of a data block is mapped into a given set of data sub-carriers and is computed from the corresponding fading amplitudes of the different sub-carriers at the receiver. Due to multi-path propagation the SINR values in this vector are received with different levels of quality and for this reason, at least theoretically, they should be considered separately by the LUT for the generation of the BLER/FER/PER/BER metrics. However, the amount of resource elements in all domains (power, time, frequency, code and space) is far too high for them to be considered individually by the LUT. As a consequence the vector of SINR values must be compressed to a lower dimension order, preferably to a one or two-dimensional vector, before being inputted to the LUT.

Assume an OFDMA based multiple access schemes with SISO, and inter-cell interference, modelled as AWGN, accumulated in the thermal noise component of the SINR. Assume also that at the nth frame interval, the radio block data symbols are mapped into a set of N data sub-carriers of the OFDM symbol. An approach for the derivation of a compressed SINR metric would be the arithmetic mean over the whole set of N sub-carriers, as given by

$$SINR_{eff}(n) = \overline{G}\frac{\left(\sum_{k=0}^{N-1} \frac{|h_k(n)|^2}{\sigma_n}\right)}{N} \tag{9.6}$$

where $\overline{G}$ is the actual geometry factor between the user, its serving and neighbouring cells, given by equation (9.11); $|h_k(n)|^2$ is the instantaneous power in the kth data sub-carrier of the sub-carrier set to which the data is mapped into in the nth frame interval; N is the size of the sub-carrier set; and σ_n is the noise power which includes interference from other cells (modelled according to Gaussian distribution).

Figure 9.7 illustrates this compression principle. The arithmetic mean-based compression scheme is the simplest and less accurate effective SINR

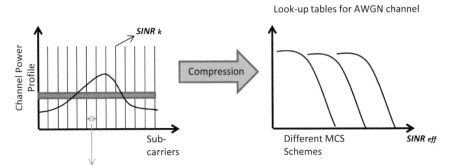

Figure 9.7 Illustration of SINR compression and mapping.

mapping. It is not accurate because it averages-out the variations of the channel along the set of sub-carriers, i.e., it underestimates some samples and overestimates others, and the inefficiencies of this approach can be seen from Figure 9.7.

9.3.3.2 Exponential Effective SINR Mapping (EESM)

The exponential ESM is derived based on the Union-Chernoff bound of error probabilities [12]. The union bound for coded binary transmission and maximum-likelihood decoding is well known and given by

$$P_e(\gamma) \leq \sum_{d=d_{\min}}^{\infty} \alpha_d P_2(d, \gamma) \tag{9.7}$$

where γ is the channel symbol SNR; d_{min} is the minimum distance of the binary code; α_d is the number of code words with hamming weight d; and $P_2(d, \gamma)$ presents the pair-wise error probability (PEP) assuming a certain Hamming distance d and a certain symbol SNR γ.

For BPSK transmission over an AWGN channel and using Chernoff-bounding techniques, the PEP can be upper bounded according to

$$P_2(d, \gamma) = Q(\sqrt{2\gamma d}) \leq e^{-\gamma d} = P_{2,Chernoff}(d, \gamma) = \left[P_{2,Chernoff}(1, \gamma) \right]^d \tag{9.8}$$

The last part of the above expression implies that the Chernoff-bounded PEP is directly given by the Chernoff-bounded (uncoded) symbol-error probability. Thus, the Chernoff-bounded error probability $P_{e,Chernoff(\gamma)}$ only depends on the weight distribution of the code and the Chernoff-bounded symbol-error probability $P_{2,Chernoff(1,\gamma)}$ according to

$$P_{ef}(\gamma) \leq \sum_{d=d_{min}}^{\infty} \alpha_d P_2(d, \gamma) \leq \sum_{d=d_{min}}^{\infty} \alpha_d \left[P_{2,Chernoff}(1, \gamma) \right]^d$$

$$= P_{e,Chernoff}(\gamma) \tag{9.9}$$

The basic principles for the Union Chernoff bound for a multi-state channel, i.e. a channel where different coded bits are subject to different SNR, is explained with the simple example of a 2-state channel. The principles are then extended in a straightforward manner to the general multi-state channel. The 2-state channel is characterized by an SNR vector $\overline{\gamma} = [\gamma_1, \gamma_2]$ where, in general, the two states γ_1 and γ_2 occur with probability p_1 and p_2 respectively. Furthermore, the two SNR values are assumed to be independent from each other, which require a corresponding interleaver in practice. Let us now look at two arbitrary code words with Hamming distance d. The SNR value, either γ_1 or γ_2, associated with each of the d differing symbols depends on the respective symbol position.

That means that the exact PEP for these two code words in case of a 2-state channel does not only depend on the distance d, but also on the position of the d differing symbols. Thus the union bound approach in the classical sense that all code-word pairs are compared would require detailed code knowledge about the bit positions. Instead, in this case the mean PEP, averaged over all possible positions of the d differing symbols, is used. This is equivalent to averaging over all possible cases on how the SNR values γ_1 and γ_2 may be distributed among the d differing symbols. Hence, the Chernoff-bounded PEP can be expressed as

$$P_{2,Chernoff}(d.[\gamma_1, \gamma_2]) = \sum_{t=0}^{d} \binom{d}{i} p_1^i p_2^{d-i} e^{-(i\gamma_1 + (d-i)\gamma_2)}$$

$$= (p_1 e^{-\gamma_1} + p_2 e^{-\gamma_2})^d \tag{9.10}$$

where the binomial theorem has been used to arrive at the final expression. To clarify the second expression, $p_1^i p_2^{d-i}$ represents the probability that i of the d differing symbols are associated with SNR γ_1 and the residual $(d-i)$ symbols are associated with SNR γ_2. There are $\binom{d}{i}$ such events and $e^{-(i\gamma_1 + (d-i)\gamma_2)}$ is the Chernoff-bounded PEP for such an event.

It can be noted that the term $p_1 e^{-\gamma_1} + p_2 e^{-\gamma_1}$ is the averaged Chernoff-bounded symbol-error probability for the 2-state channel. Therefore, the simple relationship found for the 1-state channels is also valid for the 2-state channel, i.e.

$$P_{2,Chernoff}(d.[\gamma_1, \gamma_2]) = \left[P_{2,Chernoff}(1, [\gamma_1, \gamma_2]) \right]^d \qquad (9.11)$$

Moreover, from the polynomial theorem, it can be shown that the same is true for the general multi-state channel, characterized by a vector $\bar{\gamma} = [\gamma_1, \gamma_2, \cdots \gamma_N]$, i.e. $\bar{\gamma} = [\gamma_1, \gamma_2]$

$$P_{2,Chernoff}(d.\overline{\gamma}) = \left[P_{2,Chernoff}(1, \overline{\gamma}) \right]^d \qquad (9.12)$$

This feature of the Chernoff-bounded PEP is now exploited to derive the exponential ESM. The goal is to find an effective SNR value γ_{eff} of an equivalent 1-state channel such that the Chernoff-bounded error probability equals the Chernoff-bounded error probability on the multistate channel, i.e.

$$P_{2,Chernoff}(\gamma_{eff}) = P_{2,Chernoff}(\overline{\gamma}) \qquad (9.13)$$

Due to the feature stated above, this can be achieved by matching the respective Chernoff-bounded symbol-error probabilities

$$P_{2,Chernoff}(1, \gamma_{eff}) = P_{2,Chernoff}(1, \overline{\gamma}) \qquad (9.14)$$

Inserting the Chernoff-bound expressions directly gives the exponential ESM:

$$\gamma_{eff} = -\ln\left(\sum_{k=1}^{N} p_k e^{-\gamma_k} \right) \qquad (9.15)$$

or, for the case of OFDM with N carriers and different SNR γ_k on each carrier:

$$\gamma_{eff} = -\ln\left(\frac{1}{N} \sum_{k=1}^{N} e^{-\gamma_k} \right) \qquad (9.16)$$

The above derivations have assumed binary transmission (BPSK). It is clear that, for QPSK modulation, the exponential ESM becomes

$$\gamma_{eff} = -2\ln\left(\frac{1}{N} \sum_{k=1}^{N} e^{-\frac{\gamma_k}{2}} \right) \qquad (9.17)$$

For higher-order modulation, such as 16 QAM, it is not as straightforward to determine the exact expression for the exponential ESM. The reason is that higher-order-order modulation in itself can be seen as a multi-state channel from a binary-symbol transmission point-of-view. Instead, we simply state a generalized exponential ESM including a parameter β that can be adjusted to match the ESM to a specific modulation scheme or, in the general case, a specific combination of modulation scheme and coding rate. A suitable value for the parameter β for each modulation scheme and/or coding rate of interest can then be found from link-level simulations.

$$SNR_{eesm} = -\beta \ln(\frac{1}{N} \sum_{i=1}^{N} e^{-\frac{SNR_i}{\beta}}) \tag{9.18}$$

where N is the total number of subcarriers, $SINR_i$ is a vector $[SINR_1, SINR_2, \cdots SINR_N]$ of the per-subcarrier SINR values, which are typically different in a frequency selective channel. β is the parameter to be determined for each Modulation Coding Scheme (MCS) level, and this value is used to adjust EESM function to compensate the difference between the actual BLER and the predicted BLER.

To obtain β value, several realizations of the channel have to be conducted using a given channel model (e.g., Ped B and Veh A). Then BLER for each channel realization is determined using the simulation. Using the AWGN reference curves generated previously for each MCS level, BLER values of each MCS is mapped to an AWGN equivalent SINR. These AWGN SINRs for n realizations can be represented by an n element vector $SINR_{eff}$. Using a particular β value and the vector i SINR of subcarrier SINRs, an effective SINR is computed for each realization. For n realizations, we get a vector of computed effective SINRs denoted by $SINR_{eesm}$. The goal is to find the best possible β value that minimizes the difference between computed and actual effective SINRs as shown in the following equation:

$$\beta = \arg\min_{\beta} \left\| SINR_{eff} - SINR_{eesm}(\beta) \right\| \tag{9.19}$$

9.3.3.3 Simulation Condition
- SISO
- PUSC [13] mode;
- 3GPP channel model with the velocity of 3 kmh & 60 kmh;
- 100 independent channel realizations;
- CTC with MCS formats in Tables 9.2 and 9.3;
- Ideal channel estimated is assumed.

Table 9.2 Beta values for PB channel (3 km/h).

Format	Beta (dB)
QPSK (1/2)	2.18
QPSK (3/4)	2.38
16QAM (1/2)	7.34
16QAM (3/4)	8.85
64QAM (1/2)	11.09
64QAM (3/4)	14.59

Table 9.3 Beta values for VA channel (60 km/h).

Format	Beta (dB)
QPSK (1/2)	2.12
QPSK (3/4)	2.37
16QAM (1/2)	7.53
16QAM (3/4)	8.90
64QAM (1/2)	11.01
64QAM (3/4)	14.55

Beta values of different format are trained on Pedestrian B (PB) and Vehicular A (VA) channel respectively through adequate link layer simulation of 802.16e system. The obtained beta values for look up are shown in Tables 9.2 and 9.3.

There are also alternative methods to present the verification of link layer abstraction, such as

- Effective SNR from EESM Vs. Effective SNR from AWGN for given channel realization.
- Predicted BLER Vs. Simulated BLER for given channel realization.
- Predicted BLER Vs. Simulated BLER for given effective SNR

All the verification methods above demonstrate the abstraction performance equivalently, which are presented in Figure 9.8 taking the 3GPP Vehicular channel model and QSPK modulation and 3/4 coding rate for example. In this figure the simulation results used for abstraction are the same and the samples are obtained from 100 independent channel realizations. The line curve stands for the BLER performance over AWGN channel which means the effective SNR of the line curve is the ideal theoretical value while the stars are drawn using the estimated effective SNR based on the EESM algorithm. In reverse, the performance curve over AWGN is used for look up during the abstraction procedure, so the BLER value on the blue curve stands for the predicted value and the BLER value of the red star of the same effective SNR stands for the realistic statistical BLER value from simulation. The extent of the stars

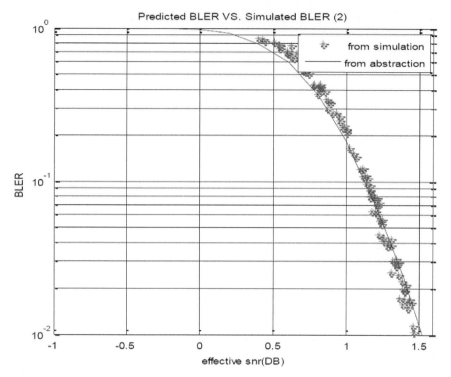

Figure 9.8 Effective SNR from EESM vs. effective SNR from AWGN.

scattering from the line shows the prediction accuracy. All the three figures above demonstrate the abstraction performance equivalently from different aspects. However, the third method is the most clear and direct way and is widely adopted in abstraction verification.

9.4 Link Layer Alternatives for Energy Efficiency

There are many aspects towards energy efficiency design in wireless devices, and must be treated in different disciplines in order to minimize the energy from a holistic perspective. At the semiconductor and transistor level, CMOS circuits energy consumption are analyzed in [14]. Taking a look at the error control processing devices as CPU, the efficiency is analyzed in terms of operations per cycle. In terms of Link Level, the objective of this chapter, we can look at the protocols related to dynamic radio resource management. Furthermore, in terms of overall system performance we look at the cellular and user behaviour in order to allocate resources efficiently, as presented in Chapter 10.

Regarding the link layer, several approaches are possible to analyze energy and efficiency improvement solutions. Even in the presented link adaptation solutions, there is extensive study towards efficient modulation and coding rate selection schemes working in synergy with ARQ. A solution that has gained much interest targets Hybrid solution aggregating forward error correction with ARQ feedback [15]. For example, one may evaluate in terms of energy, the cost of using ARQ since the time it takes to send a packet and receive confirmation creates additional overhead in terms of signalling.

9.5 Conclusions and Future Work

In this chapter we presented the link layer design of cellular wireless systems and its modelling and evaluation, including approaches for energy efficiency. The link layer architecture of new generation of wireless systems based on OFDM, namely LTE and WiMAX, and the major blocks and their interconnections were identified. We identified the dynamic characteristics of the radio resource management entity, which play an important role in the overall system efficiency. These dynamics management schemes can be used to improve energy efficiency at the link level. Moreover, we identified the link adaptation in packet based communication systems in terms of modulation and forward error coding rate selection, and the Automated Repeat Request protocol as important aspects that can be used to improve the energy efficiency of mobile wireless cellular networks. The methods of performance evaluation which are through simulation were also presented including interface to upper levels, for the specific case of OFDM systems.

Acknowledgements

The authors would like to acknowledge the project No. 23205 – GREEN-T, co-financed by the European Funds for Regional Development (FEDER) by COMPETE – Programa Operacional do Centro (PO Centro) of QREN. Valdemar Monteiro is a PhD student at Kingston University.

References

[1] 3GPP, Multiplexing and channel coding FDD (release 9). 3rd Generation Partnership Project, 3GPP TS 25.212, Tech. Rep., March 2012.

[2] 3GPP, A global initiative, http://www.3gpp.org/.

[3] 3GPP, HSPA, High-Speed Packet Access,, http://www.3gpp.org/HSPA.

[4] H. Holma and A. Toskala, *WCDMA for UMTS – HSPA Evolution and LTE*, 4th ed. Wiley, 2007.

[5] 3GPP, Performance comparison of bit level and symbol level chase combining. 3rd generation partnership project, Texas Instruments TSG-RAN Working Group 1 3GPP TS meeting #20 Busan, Korea, Tech. Rep., May 21–25 2012.

[6] V. Hajovska et al., Harq schemes for HSDPA – Analysis and simulation, in *Proceedings 50th International Symposium ELMAR (ELMAR-2008)*, September 2008.

[7] D. Chase, Code combining: A maximum-likelihood decoding approach for combining an arbitrary number of noisy packets, *IEEE Trans. on Commun*, vol. 33, pp. 593–607, May 1985.

[8] T. Chen, H. Kim, and Y. Yang, A novel interface between link and system level simulations, in *Proceedings of ACTS Summit 1997*, October 1997.

[9] K. Brueninghaus, D. Astely, T. Salzer, S. Visuri, A. Alexiou, S. Karger, and G.-A. Seraji, Link performance models for system level simulations of broadband radio access systems, in *Proceedings of IEEE 16th International Symposium on Personal, Indoor and Mobile Radio Communications (PIMRC2005)*, vol. 4, pp. 2306–2311, September 2005.

[10] IEEE Standard 802.16e-2005, Amendment to IEEE Standard for Local and Metropolitan Area Networks – Part 16: Air Interface for Fixed Broadband Wireless Access Systems-Physical and Medium Access Control Layers for Combined Fixed and Mobile Operation in Licensed Bands, 2005.

[11] E. Tuomaala and H. Wang, Effective SINR approach of link to system mapping in ofdm/multi-carrier mobile network, in *Proceedings of 2nd International Conference on Mobile Technology, Applications and Systems*, pp. 5, November 2005.

[12] 3GPP-TSG-RAN WG1 #35, System-level evaluation of OFDM – Further considerations, (Rl-031303) November 17–21 2003 Lisbon, Portugal, Tech. Rep., November 2003.

[13] WiMAX Forum, WiMAX system evaluation methodology. Version 1.0, January 2007.

[14] A. P. Chandrakasan and R. W. Brodersen, Minimizing power consumption in digital cmos circuits, *Proceedings of the IEEE*, vol. 83, no. 4, pp. 498–523, April 1995.

[15] P. Lettieri, C. Schurgers, and M. B. Srivastava, Adaptive link layer strategies for energy efficient wireless networking, *Wireless Networks*, vol. 5, pp. 339–355, 1999.

10

System Level Evaluation Methodology for Energy Saving

Valdemar Monteiro[1,2], Shahid Mumtaz[1], Alberto Nascimento[3], Jonathan Rodriguez[1], and Christos Politis[2]

[1]*Instituto de Telecomunições, Aveiro, Portugal*
[2]*University of Kingston, Kingston upon Thames, UK*
[3]*Universidade da Madeira*
e-mail: vmonteiro@av.it.pt

Abstract

The growing demand in data traffic coupled with the drive towards new generation handsets that are more power hungry are providing the impetus for new research initiatives towards energy saving. Although much research in energy efficiency has been geared towards hardware power optimization, the energy consumption of new generation wireless (3G and 4G) is sensitive on the design and flexibility of the dynamic resource allocation protocol, with a high fraction of the holistic power consumption attributed to the data transmission over the wireless connection. This sensitivity suggests that there is an opportunity for optimization of the energy consumption based on communications protocols, namely the link and transport layer protocols. In this chapter, we extend the analysis given in the prequel to understand how the link level performance interacts with the system level to form part of the system level model for analyzing system performance in cellular networks. Moreover, this chapter focuses on the dynamic resource allocation entity which is responsible for managing radio resources in wireless systems. We identify the energy saving performance metrics and provide a holistic system level performance analysis using traditional scheduling policies from an

Shahid Mumtaz and Jonathan Rodriguez (Eds.), Green Communication in 4G Wireless Systems, 203–257.

energy saving perspective. This reveals that the scheduling policy plays an important role in reducing the energy consumption, highlighting that there is still much research to undertake on this area.

Keywords: Simulation, cross-layer, system-level, multi-RAT, beyond 3G.

10.1 Introduction

In the previous chapter we presented the link level approach where communications techniques are used to improve link capabilities, like OFDM. In this chapter, a system level evaluation approach is taken taking into account interference multi-links. Moreover, the system level evaluation takes into account the advantage of multi-user diversity both in terms of traffic demand and radio channel conditions for protocols and related algorithm enhancements and the performance results are given in terms of overall system performance.

Several approaches can be taken for system level evaluation where there is always a trade-off between maximizing throughput, ensuring fairness between served users, complexity and accuracy. The choice of the right simulator category depends on the investigated problem and the desired results. Usually, it is advantageous to decouple simulators working at different levels of abstraction in order to reduce complexity. In this case, appropriate interfaces are required between the simulators to provide a service point to the connecting layers. Since System Level Simulation (SLS) are often confused with system emulators and/or link layer simulators, we will clarify their main differences, including the objectives and common functionalities. The first step in the design of a simulator or an emulator is to define the system to be analyzed. In this step it is necessary to determine the input data or initial conditions of the system, the system parameters or state variables and the performance metrics or output data of the system. In addition, their characteristics vary according to the type of system to be simulated or emulated. In general, a system simulator attempts to recreate the same system parameters by using abstract theoretical models. It also aims at collecting performance metrics to assess different properties of the system.

In contrast, an emulator does not aim at collecting performance metrics, but only at recreating system parameters. In general, an emulator dependency on abstract models is not as tight as that of a system simulator. In many cases an emulator is a piece of hardware that exactly reproduces all the desired system parameters.

This chapter is organized as follows: in Section 10.2 we elaborate the theme of energy efficiency and associated challenges for radio resource management; in Section 10.3 the methodology for system level simulations is presented, including the requirements for system modeling; in Section 10.4 we present the energy efficiency for scheduling protocols, we include the analysis of MIMO vs. SISO systems; and in Section 10.5 we showcase other approaches given in the literature for system level simulations.

10.2 Radio Resource Management for Energy Efficiency

Fundamental principles of energy efficient Radio Resource Management (RRM) will be explained in this section. Techniques involving dynamic resource allocation and other algorithms involving MAC layer protocols, like scheduling and link adaptation, can be used to improve the efficiency in terms of energy consumption. In this section we present fundamental principles for energy efficient RRM.

10.2.1 Energy Efficiency Metrics in Wireless Systems

Energy metrics are very important to evaluate the Efficiency of MAC-PHY protocols in next generation wireless systems. We have come to understand several energy metrics which are used for benchmarking in terms of energy efficiency or saving. An extended overview of energy metrics for green wireless communications is described in [14]. Energy efficiency metrics are categorized according to their use-case scenario [14]:

- to compare the energy consumption performance of different components and systems in the same class.
- to adjust the target on energy for long term research and development basis.

To investigate energy-aware networking topologies energy metrics are used most cases for maximizing the energy efficiency of the system, a term that is now being commonly adopted in the same light as spectral efficiency. An important energy metric to be considered is the energy efficiency (EE) of the system, which is the ratio of the total transmitted bits per unit energy consumption (unit: bits/Joule):

$$E = \frac{\text{Transmittted bits}}{\text{Energy}} \left[\frac{\text{bits/Second}}{\text{Watt}} = \frac{\text{bits}}{\text{Second} \times \text{Watt}} = \text{bits/Joule} \right]$$

(10.1)

Moreover, an energy metric which is also being considered is the energy consumption ratio (ECR) [8] which is the energy per transmitted information bit (unit: Joule/bits):

$$\text{ECR} = \frac{\text{Energy}}{\text{Transmittted bits}} \left[\frac{\text{Watt}}{\text{bits/Second}} = \frac{\text{Second} \times \text{Watt}}{\text{bits}} = \text{Joule}/\text{bits} \right]$$

$$(10.2)$$

Energy metrics can also be divided in three levels: component level, equipment level and system/network level. Energy metrics for component and equipment level have already been well established [14]. System level energy metrics still have to be consolidated to establish a permanent benchmark criterion. In this chapter we focus on the system level for cellular networks.

Some basic energy metrics for cellular networks are given by the amount of coverage area, as described in the following:

- In rural areas, the EE metric is defined in [14] as

$$PI_{\text{rural}} = \frac{A_{\text{coverage}}}{P_{\text{site}}} \qquad (10.3)$$

  where A_{coverage} is the Radio Base Station (RBS) coverage area in km^2, and P_{site} is the average site power consumption. The metric for urban area is defined as:

$$PI_{\text{urban}} = \frac{N_{\text{BH}}}{P_{\text{site}}} \qquad (10.4)$$

  where N_{BH} is the number of subscribers based on the average busy hour traffic demand by subscribers and on the RBS busy hour traffic.
- The metric for the area power consumption is defined in [16] as

$$\rho = \frac{P_C}{A_C} \qquad (10.5)$$

  where PC is the power consumption of a cell site, and AC is the coverage area in km^2 of the cell.

We should keep in mind that the aforementioned metrics are not directly related to the throughput performance of the system since in GSM and WCDMA systems, the main service is voice, the performance of which is not measured by the data rate. Nevertheless, in fourth generation (4G) cellular systems, packets are used to carry all services. This opens the way to measure the performance of the system by throughput represented by the data rate. Bit/Joule is expected as the basic EE metric for fourth generation cellular systems and beyond.

Table 10.1 Linear power model parameters [29].

Parameter	Description
N_{sector}	Number of sectors
P_{TX}	Transmit power
P_{SP}	Signal processing overhead
C_{PSBB}	Battery backup and power supply loss
N_{PApSec}	Number of power amplifiers per sector
η_{PA}	Power amplifier efficiency
C_C	Cooling loss

10.2.2 Power Consumption Models for Cellular Networks

The power consumption model provides the energy consumption of different types of cellular networks. Usually micro cells are more energy efficient than macro cells wireless systems due to the path loss exponent factor. The power consumption model for 4G network described in [29] can be used as a stepping stone for simulating the energy performance of the network, where basic power related parameters of the network are summarized in Table 10.1 among others. Additional parameters exist and are fixed for each simulation run, these can include the technology and environmental conditions which are important for cooling, etc.

The power consumption of a micro base station consists of two parts. The first part describes the static power consumption, a standard power figure which is consumed even when the base station has no load. Depending on the load situation, a dynamic power consumption component is added to the static power. Arnold et al. [29] derived a power model for typical base stations as deployed in todays networks. These provide a relative small dynamic contribution to the power consumption and the optimum cell size is strongly affected by the static part. The macro base station power consumption model is given by [29]

$$P_{\text{BS,Macro}} = N_{\text{sector}} \cdot N_{\text{PApSec}} \cdot \left(\frac{P_{\text{TX}}}{\eta_{\text{PA}}} + P_{\text{SP}} \right) \cdot (1 + C_C) \cdot (1 + C_{\text{PSBB}}) \quad (10.6)$$

For a micro base station, the power consumption model is more dynamic. The power model constitutes two parts which is given by

$$P_{\text{BS,Micro}} = P_{\text{static,Micro}} + P_{\text{dynamic,Micro}} \quad (10.7)$$

The parameters for micro base station power model parameters are listed in Table 10.2.

Table 10.2 Micro base station power model parameters [29].

Parameter	Description
$P_{static,Micro}$	Static power consumption
P_{TX}	Maximum transmit power per power amplifier
N_L	Numbers of active links
$C_{TX,NL}$	Dynamic transmit power per link
$P_{SP,NL}$	Dynamic Signal processing per link
$P_{dynamic,Micro}$	Dynamic power consumption
η_{PA}	Power amplifier efficiency
$C_{TX,static}$	Static transmit power
$P_{SP,static}$	Static signal processing
C_{PS}	Power supply loss

The static power consumption for the micro base station is given by [29]

$$P_{static,Micro} = \left(\frac{P_{TX}}{\eta_{PA}} C_{TX,Static} + P_{SP,static} \right) \cdot (1 + C_{PS}) \qquad (10.8)$$

And for the dynamic power consumption component [19]:

$$P_{dynamic,Micro} = \left(\frac{P_{TX}}{\eta_{PA}} (1 - C_{TX,static}) \cdot C_{TX,NL} + P_{SP,NL} \right) \cdot N_L \cdot (1 + C_{PS})$$
$$(10.9)$$

However, the components with the highest effect on a base station's power consumption are the following [19]: utilization of remote radio heads or ordinary power amplifiers with corresponding feeder losses, different kinds of cooling (air conditioning, air circulation, or free cooling), site sharing (especially regarding infrastructure), and number of carrier frequencies. A key approach that operators are currently exploring as a concerted effort towards energy saving is network sharing, which is perceived to have the most impact towards reducing the operators capital and operational expenditure in the network.

10.3 System Level Simulation Methodology

In this section we present the methodology for System Level Simulation incorporating energy efficiency metrics. We focus on OFDM cellular and typical environments namely urban and rural, including accurate radio channel and interference modeling.

Figure 10.1 shows a system with multiple cells where in each cell there is one base station and a random number of mobiles, randomly located inside

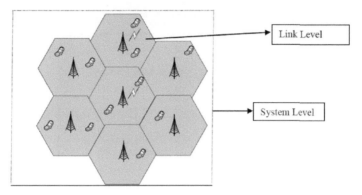

Figure 10.1 Wireless system level simulation.

each cell. Mobiles access the network through the Base Station (BS). A comprehensive system level tool is needed to evaluate the performance of such systems, which captures every aspect of the real cellular environment. This kind of tool is called a System Level Simulator (SLS). A single simulator approach that constitutes the two approaches, as presented in Figure 10.1, including the multi-link level and radio resource management protocols would be preferred. However, the complexity of such a simulator is far too high with the required simulation resolutions and simulation processing times. In addition, the scope of both simulators are very distinct, i.e. a link level simulates a single link, one base station and one user whereas a system level simulator simulates multiple links as shown in Figure 10.1.

At the link level, as presented in the previous section, the granularity is in order of the BER, or if it is a CDMA/WCDMA system, it can be as low as nanoseconds according to the respective chip rate. While at the network level, it is at the level of packet durations, which is typically several orders of magnitude higher than the BER. Due to these computational complexities, it is not possible to simulate all networks in one system level simulator. Therefore, separate link and system level simulations are required with an appropriate methodology to pass the results from the link level to the system level. This is the so-called Link-to-System (L2S) interface as shown in Figure 10.2, which was discussed in more detail in Chapter 9. In practice, this interface is realized through a set of mapping tables known as Look up Tables (LUT). These mapping tables are constructed at the link level and they represent a tabulated BER or FER function as a function instantaneous system level SINR.

Figure 10.2 Interface between system level and Link *l*.

10.3.1 Requirements for System Level Simulation Modeling

Implementation of a system level tool requires complex modeling of the real environment. As shown in Figure 10.3, several aspects need accurate modeling. The minimum modeling of wireless systems should involve three aspects:

- Users dynamic behavior, namely user mobility;
- Traffic characteristics for the applications considered;
- Radio aspects involved in the transmission of the signals from users to destination.

Taking a deeper look into the characteristics of simulation environment, we need to capture certain traits which pertain to real-life instances in cellular networks, therefore, when considering the users behavior, as their position can be random, there is need for a deployment model. Secondly, as they are mobile, there is need for a mobility model related to each scenario or environment. Thirdly, as the user uses wireless communication protocols to transport applications or services there is need for a model of the data traffic. Finally, the information is sent by a radio waveform modulated by a channel, thus creating the need to model the impact of the channel on the signal quality requiring appropriate propagation models. Therefore, the main modeling requirements for the SLS, as shown in Figure 10.3, are:

- Deployment Modeling;
- Mobility Modeling;
- Traffic Modeling;
- Radio Channel Propagation Modeling;
- Transmission/reception techniques.

The methodology followed in the system level simulations depends on a different set of assumptions regarding: type of wireless system simulated; air interface technology; simulation complexity and time resolution; interface with other layers of the protocol stack, such as the physical layer and/or network layer; channel and interference modeling and application traffic models. In particular, the following aspects must be considered when

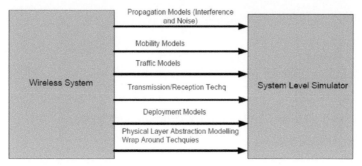

Figure 10.3 Modeling of a wireless communication system.

developing a system level tool and performing system level simulations:

Network scenario

- This is related to the particular environment considered in the simulations: urban, rural, vehicular or indoor. Each has a specific user mobility pattern.

Network layout

- Amount of tiers and number of base stations simulated. This is crucial for the amount of interference considered.
- Type of cells in each site/base station: one omnidirectional cell or three, six sectored cells for example.
- Number of mobile stations and their distribution over the network coverage area.

Radio resource management

- Power control.
- User mobility and handover.
- Definition of the radio resources according to the type of air interface and the medium access layer.

Physical layer modeling and abstraction

- Definition of the metrics used to map the physical layer performance to higher layers of the protocol stack.
- Type of interface used in the interaction between system and physical layers.

Propagation and channel modeling

- Path loss propagation.
- Slow fading (shadowing) propagation.
- Fast fading channel modeling.

Interference modeling

- Intra-cell and inter-cell and inter-system interference.

Implemented radio access system

- Multiple access to radio resources, circuit switch/packet switch.

Traffic models for application services

- Choice of traffic models: emulation by using pre-defined traffic models or use of real traces from real networks.

Performance metrics

- Metrics for network evaluation performance.
- Metrics for user satisfaction evaluation.

10.3.2 Simulation Execution Flow

Two different types of simulations can be performed at system level: using a Combined Snapshot-Dynamic or a Dynamic mode.

- Dynamic mode: in this mode fully dynamics are simulated. Mobility is enabled as mobiles travel along the network coverage area performing handovers. Mobiles are dropped in the network in the beginning of the simulation run and remain active since the instant of activation, which can be coincident with the beginning of the simulation run or be defined by some random distribution. Only one simulation run is performed and mobiles are removed at the end of the simulation. Statistics are collected as mobiles travel through the network coverage area. All channels components, including path loss, shadowing and fast fading propagation components are re-computed at every transmission time interval. The new position of the mobile station in the next transmission time interval is also computed according to the chosen mobility model. The trade-off of this mode is the inherent complexity, since all channel components are calculated in each iteration.
- Combined snapshot-dynamic mode: mobility and handovers are disabled in this mode and a given number of simulation runs are performed. Mobile stations are drawn on the network in the beginning of each

simulation run and are removed at their end. They remain active since the instant of activation, which can be coincident with the beginning of the simulation run or be defined by some random distribution. In this mode, path loss and shadowing are computed at the beginning of each simulation run and remain constant until the end of the run. Fast fading is re-computed at every transmission time interval. This mode increases the simulation speed as the different simulation runs (snapshots) can be performed in parallel. The final results should be averaged on the number of snapshots.

10.3.3 Simulation Methodology for Evaluating Scheduling

In both modes mobiles are randomly uniformly distributed over the hexagonal cells. Each base station can be configured with one sector (omnidirectional antenna pattern) or with three sectors/cells (directional antenna pattern). However when evaluating MAC protocols like Scheduling, the combined snapshot-dynamic mode can be a valuable method to conduct system level simulations. A meaningful number of simulation runs (snapshots) should be performed along each evaluation. A Complementary method to reduce the complexity and thus the processing time during the simulation run is to consider mobile stations in the first tier of cells only. The neighboring cells contribute only to interference generation. With this approach the number of mobiles are reduced as well as the number of channel calculations whilst ensuring sufficient accuracy for evaluating the impact of scheduling policies in MAC protocols.

In this method path loss and shadowing are computed at the beginning of each run and for each mobile-base station pair (including neighboring cells), and are kept constant until the end of the run; whilst fast fading is executed for each transmission time interval. The steps followed in the simulation flow for a single run of a general system level simulation tool are as follows, for both omni or sectorized base stations:

- Mobile stations are dropped independently, with uniform distribution throughout the system. Each mobile corresponds to an active user session that runs for the whole run.
- Mobiles are assigned channel models. This can be a channel mix or separate statistical realizations of a single type of channel model.
- Mobiles are assigned a traffic model and packets are generated according to the desired traffic model/service.

- Cell assignment is based on the received power at the mobile station from all potential serving cells. The cell with the best path to the mobile station, taking into account slow fading, path loss and antenna gains, is chosen as the serving sector.
- For simulations that do not involve handover performance, evaluation of the location of each mobile station remains unchanged during a drop and the mobiles speed is used only to determine the Doppler effect of fast fading. The mobile station is assumed to remain attached to the same base station for the duration of the drop.
- For a given drop the simulation is run for the pre-defined duration and then the process is repeated with the mobile stations being dropped at new random locations.
- Performance statistics are collected for mobile stations in all cells.

Each run is made up of a number of transmission time intervals or frame periods, as each transmission time interval lasts for the period of time equal to the transmission of a single OFDM frame. Regarding execution flow, the simulator developed for B3G/4G standards LTE, WiMAX or HSPA system level simulations is basically a finite machine whose states repeat at each transmission time interval. A transmission time interval (TTI) should be equal to frame period. In our approaches the TTI for LTE is 1ms, for WiMAX is 5 ms and for HSPA is 2 ms. For each TTI the following events are performed:

- Fast fading is computed for each mobile station in each transmission time interval. Slow fading and path loss are assumed as constant during the whole simulation run.
- Packets are withdrawn from buffers assigned to traffic models. Packets are not blocked, as the queues are assumed as infinite. Start times for each traffic type for each user should be randomized.
- Map of radio resources allocation is updated.
- Packets are scheduled with a packet scheduler using the required metric. Packet decoding errors result in packet retransmissions. In the Dynamic Resource Allocation (DRA) module a Hybrid Automatic Repeat Request (HARQ) process is modeled by explicitly rescheduling a packet as part of the current packet call and after a specified feedback delay period.
- The map of radio resources allocation is computed, according to the implemented scheduler. This is performed by the Dynamic Resource Allocation (DRA) Module.
- Packets are transmitted.

Table 10.3 LTE Transport block size and bit rate associated to MCS.

MCS	Modulation, code rate	Transport block size [bits]	R(CQI) [Mbps]
MCS 1	QPSK 1/2	4200	4.82
MCS 2	QPSK 3/4	6300	6.3
MCS 3	16-QAM 1/2	16800	16.8
MCS 4	16-QAM 3/4	25200	25.2
MCS 5	64-QAM 1/2	25200	25.2
MCS 6	64-QAM 3/4	37800	37.8

- Packet quality detection is performed and feedback regarding the status of the decoding is reported back.

10.3.4 Resource Block Definition

Resource block (RB) is the basic unit that contains data to be transmitted to the mobile users in each TTI. In terms of system level simulation, RB is seen as the amount of data unit that is transmitted to the mobile user. In the specific case of OFDM based systems, like LTE or WiMAX, RB represents amount of data carried by sub-carriers and time slots within a frame. Figure 10.4 presents the RB in the frame structure of the LTE standard. While the number of slots and sub-carriers are fixed in each RB, the amount of data varies according the modulation and coding rate (MCS), which is the link adaptation method for channel status. The allocation of the proper sub-channels should be able to exploit physical layer information such as SINR, MCS level and velocity of the user, attempting to maximize resource efficiency and constrained to the power available for data transmission at the base station. The velocity is very important because it determines the proper type of sub-channelization mode used (adjacent AMC or diversity permutation). Table 10.3 presents the resource block size in bits, the modulation used (MCS) and respective achievable bit rate used by LTE system simulator.

10.3.5 Simulation of Packet/Data Decoding Process

The Block Error Rate (BLER) resulting from decoding the information transmitted along a single Resource Unit (RU) is denoted by $BLER_{RU}(SINR_{RU})$ and is obtained from the link-to-system interface between the PHY and MAC layers, using as input the Signal to Interference plus Noise Ratio, $SINR_{RU}$. Then a random variable, uniformly distributed between 0 and 1, is drawn. In case the random value is less than $BLER_{RU}(SINR_{RU})$ the block is considered

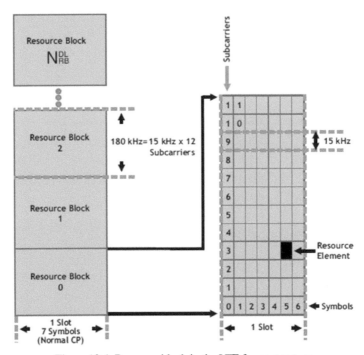

Figure 10.4 Resource block in the LTE frame structure.

as erroneous and a Negative Acknowledge (NACK) message is sent back to the base station on the associated signaling channel. Otherwise, the block is deemed as error free and an Acknowledge (ACK) message is transmitted.

The success or failure in the decoding of the transmitted block of information is computed from decoding each individual resource unit into which the data block is mapped. Assuming that a total amount of N_{res} radio resources are used in the transmission and that the decoding is an independent and identically distributed random process, the BLER for the whole radio block is given by equation 10.10.

$$BLER_{RB} = 1 - [1 - BLER_{RU} (SINR_{RU})]^{N_{res}} \qquad (10.10)$$

10.3.6 Simulation Complexity and Time Resolution

There is a trade-off between accuracy and complexity resulting in simulation execution time. The correct balance must be found for each aspect of the system to be evaluated, and in this chapter we will not say which is best.

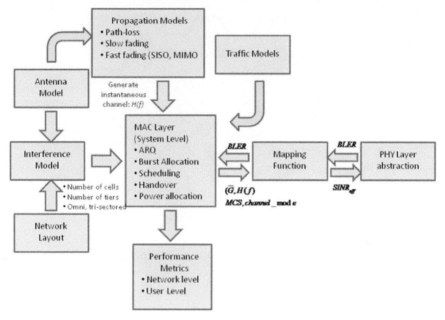

Figure 10.5 Simulation components.

We will rather give the guidelines and approaches usually taken for typical cases. The main components of a complete system level simulation tool and their interconnection and flow are illustrated in Figure 10.5, according to simulation procedures elaborated in [19, 21].

10.3.7 Network Scenario and Layout

All system level simulations conducted in this work were performed assuming an urban environment scenario. The simulated network is constituted of 57 sectors (19 base stations with three sectors each), composing a 3-tier hexagonal cellular network layout, as illustrated in Figure 10.6. The antenna pattern used in each sector is plotted in Figure 10.7. The antenna pattern used for the sectored antenna deployment only considers the horizontal pattern, corresponding to a main sector of 70 degrees. According to the model for the typical antenna pattern proposed in [15], power attenuation is computed as a function of the angle between the antenna pointing direction and the mobile

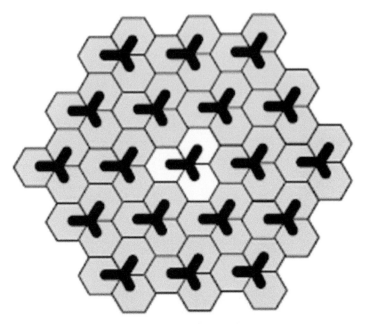

Figure 10.6 Network layout deployment.

3 Sector Antenna Pattern

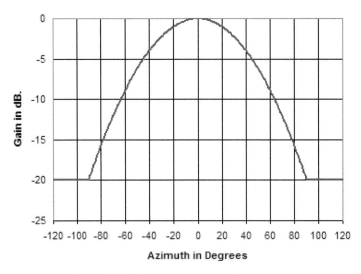

Figure 10.7 Antenna pattern for three sectors.

to base station direction, as given by

$$A(\theta) = -\min\left[12\left(\frac{\theta}{\theta_{3\text{dB}}}\right), A_m\right] \qquad (10.11)$$

where $-180 < \theta < 180$ is the angle between the antenna's pointing direction and the mobile to base station line-of-sight direction in degrees; $\theta_{3\text{dB}} = 70$ is the beam width at 3 dB; and $A_m = 20$ dB is the maximum attenuation. Two types of cell configurations can be defined for simulations: central-cell and non-central cell approach. In the central-cell approach mobiles are dropped along the coverage of the central base station and statistics are collected only for the cells of this base station. Naturally the central cell approach simulation method can be enabled only in conjunction with the combined snapshot-dynamic mode, as mobility modeling is disabled. The cells in the remaining tiers are assumed as fully loaded, i.e., transmitting with maximum power and contribute to interference only.

10.3.8 Propagation Channels Models

The radio propagation is divided into three distinct components, namely path loss, slow fading (shadowing) and fast fading. The decrease of the transmitted radio signal impinging on the receiver antennas is the result of their contribution. Accurate modeling of each one of these three radio propagation components depends on the simulation scenario envisaged for the system-level simulations. Namely the simulation scenario can be described according to the following characteristics:

- Type of environments: indoor, urban, suburban and rural.
- Mobile speed: pedestrian, vehicular, train.
- Type of receiver used in the signal processing at the receiving end.
- Antenna radiation pattern.
- Antenna configuration used in the communication between the transmitter and the receiver (SISO, SIMO, MISO, MIMO).
- Radio transmission parameters: carrier frequency, system bandwidth, etc.

10.3.8.1 Path Loss Model

Path loss is defined as the power loss due to the propagation environment. The signal attenuation is directly proportional to a power of the distance between the transmitter and the receiver. The attenuation also depends on the carrier frequency and on the type of environment. The model used for the computation of the attenuation of the radio signal between the transmitter and the receiver is the one proposed in [15] for vehicular environments. This model is suitable for both urban and suburban scenarios, in which the buildings form a relatively homogenous clutter. According to Arnold et al. [29] the path loss in dB is given by

$$L_{[dB]} = \left[40 \left(1 - 4 \times 10^{-3} \frac{\Delta h_b}{m} \right) \right] \log_{10} \left(\frac{R}{Km} \right)$$
$$- 18 \log_{10} \left(\frac{\Delta h_b}{m} \right) + 21 \log_{10} \left(\frac{f}{MHz} \right) + 80 \text{ dB} \quad (10.12)$$

where R represents the distance in kilometers between the base station and the mobile; f is the carrier frequency; and Δh_b is the base station antenna height from the roof level. Simulations were performed assuming the carrier frequency proposed in WiMAX profile, which is equal to 2.5 GHz. Also, it was assumed an antenna height, Δh_b, at the base station equal to 15 m. For this setting, the path loss in equation (10.12) results in the formula presented in equation (10.13), which is the expression used for the computation of the path loss in the simulations:

$$L_{[dB]} = 130.18 + 37.6 \log_{10} \left(\frac{R}{Km} \right) \quad (10.13)$$

10.3.8.2 Shadowing (Slow Fading) Model

Shadowing is the slow variation of the signal power at the receiver. It is given in dB and is modeled by a Gaussian random variable with linear autocorrelation, which is an exponential function of the de-correlation distance, d_{corr}, according to Gudmundson [22] and is given by

$$\rho(d) = e^{-\ln(2) \frac{d}{d_{corr}}} \quad (10.14)$$

The parameter d_{corr} is the length of the de-correlation distance for which the auto-correlation of the shadowing process, ρ, is equal to 0.5.

Although Gaussian random processes can be modeled as a sum of sinusoids (SOS), conventional one-dimensional channel models (1-D) cannot capture the spatial correlation of shadowing processes. For example, when a given mobile is moving along a closed path around its base station, 1-D models cannot capture the influence of the slow shadowing, in the variation of the signal received at the mobile station, and this affects the performance

Table 10.4 Parameters for shadow fading model.

Parameter	Values
Log-Normal Shadowing std σ_{SH}	8 dB
De-correlation length d_{corr}	20 m

of the handover algorithm. According to this, Cai et al. [12] propose a two-dimensional (2-D) SOS-based channel model to simulate slow fading. The shadowing $SH^j_{(x,y)}$ in dB between one mobile station at position (x, y) and base station j is the sum of two spatial functions, F_0 and F_j, having a Gaussian distribution, with standard deviation mean equal to σ_{SH} (the shadowing standard deviation in dB) and auto-correlation given by (10.13), using the method described in [19]. It is given by

$$SH^j_{(x,y)} = \sqrt{0.5}x \left[F_o(x, y) + F_j(x, y) \right] \qquad (10.15)$$

The standard deviation, σ_{SH}, and the de-correlation length, d_{corr}, for the urban scenario used in the system level simulations are the ones listed in Table 10.4.

10.3.8.3 Fast Fading Model

The fast fading component of the signal is simulated by fast generation of independent Rayleigh faders, according to a modified Jakes model from the method proposed in [26, 27]. In order to speed-up simulations, the multi-path channel model is used for the serving cell, while a flat fading channel model (with only one tap) is assumed for neighboring cells. The mobile speed and carrier frequency are the parameters considered in the generation of the fading statistics. In this context a channel model corresponds to a specific number of paths, a power profile giving the relative powers of these multiple paths and Doppler frequencies to specify the fade rate.

ITU multi-path channel models for narrow-band SISO are proposed in [24]. These models are based on a discrete version of the scattering function of the propagation channel and are designated as tapped delay line models. Each tap is characterized by an attenuation, A_i, a corresponding delay τ_i, a Doppler frequency f_d, and a Doppler Power Spectrum (DPS) $P_s(f_d, \tau_i)$, at the ith tap. A separate link level simulation must be performed for each specific channel model and mobile stations' velocity combination.

Tables 10.5 and 10.6 detail the parameters used in the definition of each type of channel model proposed by ITU [3]. The channel model assigned to a specific user remains fixed over the whole duration of a simulation run.

In [18] system level simulations are conducted to validate Mobile WiMAX standard using the ITUs normalized power profiles for channel mod-

Table 10.5 Parameters for the different types of fast fading channel models for SISO.

Channel model	Multi-path model	Number of Paths	Speed (Km/h)	Fading
Model 1	Ch-100	1	30	Jakes
Model 2	Ch-100	1	120	Jakes
Model 3	Ch-104	6	30	Jakes
Model 4	Ch-104	6	120	Jakes
Model 5	Ch-102	4	3	Jakes
Model 6	Ch-103	6	3	Jakes

Table 10.6 Multi-path channel models for performance simulation.

Channel model		Path 1	Path 2	Path 3	Path 4	Path 5	Path 6
Flat Fading Ch-100	Path Power (dB)	0	–	–	–	–	–
ITU Ped. A Ch-102	Path Power (dB)	–0.51	–10.21	–19.71	–23.31	–	–
	Delay (ns)	0	110	190	410	–	–
ITU Ped. B Ch-103	Path Power (dB)	–3.92	–4.82	–8.82	–11.92	–11.72	–27.82
	Delay (ns)	0	200	800	1200	2300	3700
ITU Veh. A Ch-104	Path Power (dB)	–3.14	–4.14	–12.14	–13.14	–18.14	–23.14
	Delay (ns)	0	310	710	1090	1730	2510

els such as ITU Vehicular A: Ch-104 and ITU Pedestrian-B: Ch-103 [3, 25]. These models are illustrated in Table 10.6. The absolute power values are normalized so that they sum to zero dB (unit energy) for each given channel.

10.3.8.4 MIMO Channel Modeling in System Level Simulations

The MIMO channel modeling for the system level simulations is presented in Appendix B.

10.3.9 Signal to Interference Plus Noise Ratio (SINR) Modeling

In system level simulations mobile stations are randomly dropped along the simulated coverage area. When the mobile station becomes active, its serving cell is selected according to signal strength and the mobile station camps on this cell. As mentioned in the previous sections, the signal coming from the serving cell is modeled as a frequency selective fading channel, whereas the signal coming from neighboring cells is modeled according to a flat frequency fading channel. Assume then a given mobile station MS_i is camping in the coverage area of cell $Cell_i$. Assume also a SISO channel. The power received from serving base station $BS_{serving}$ for data sub-carrier i, $i \in [0, \ldots, N_{data} - 1]$ on mobile station MS_i in the nth frame interval is given by

$$P^{(i)}_{\text{BS}_{\text{serving}}}(n) = \frac{P^{(i)}_{\text{data}} \cdot \left| H^{(i)}_{\text{BS}_{\text{serving}}}(n) \right|^2 G_{\text{BS}_{\text{serving}}} G_{\text{MS}_i}}{\text{PL}_{\text{MS}_i\text{BS}_{\text{serving}}} \text{SH}_{\text{MS}_i B_{\text{serving}}} L_{\text{loss}}} \qquad (10.16)$$

where

- $|H^{(i)}_{\text{BS}_{\text{serving}}}(n)|^2$ is the instantaneous power from the serving base station $\text{BS}_{\text{serving}}$ at the ith data sub-carrier at the nth frame interval;
- G_{MS_i} is the gain of the antenna at the mobile station MS_i;
- $G_{\text{BS}_{\text{serving}}}$ is the gain of the antenna at the serving base station $\text{BS}_{\text{serving}}$;
- $\text{PL}_{\text{MS}_i\text{BS}_{\text{serving}}}$ is the path loss between serving base station $\text{BS}_{\text{serving}}$ and mobile station MS_i;
- $\text{SH}_{\text{MS}_i\text{BS}_{\text{serving}}}$ is the shadowing loss between serving base station $\text{BS}_{\text{serving}}$ and mobile station MS_i; and
- L_{Loss} encompasses the other losses in the transmission (cable losses, body loss, etc.).

As sub-carriers are mutual exclusively assigned inside each cell there is no intra-cell interference. Therefore, only inter-cell interference must be considered. The interfering power arriving at mobile station MS_i from neighboring cells is given by

$$P^{(i)}_{\text{Inter}}(n) = \sum_{\text{BS}_j \in \{\text{BS}_{\text{Inter}}\}} \frac{P^{(i)}_{\text{data,BS}_j} \cdot |H^{(i)}_{\text{BS}_j}(n)|^2 G_{\text{BS}_j} G_{\text{MS}_i}}{\text{PL}_{\text{MS}_i\text{BS}_j} \text{SH}_{\text{MS}_i\text{BS}_j} L_{\text{Loss}}} \qquad (10.17)$$

where

- $\{B_{\text{Inter}}\}$ is the set of interfering base stations and $\text{BS}_j \in \{B_{\text{Inter}}\}$;
- $|H^{(i)}_{\text{BS}_j}(n)|^2$ is the instantaneous power from the interfering base station BS_j at the ith data sub-carrier at the nth frame interval;
- G_{BS_j} is the gain of the antenna at the interfering base station BS_j;
- $\text{PL}_{\text{MS}_i\text{BS}_j}$ is the path loss between interfering base station BS_j and mobile station MS_i;
- $\text{SH}_{\text{MS}_i\text{BS}_j}$ is the shadowing loss between interfering base station BS_j and mobile station MS_i; and
- L_{Loss} encompasses the other losses in the transmission (cable losses, body loss, etc.).

According to equations (10.16) and (10.17), the SINR at sub-carrier i and for the nth frame interval is given by

$$\text{SINR}^{(i)}(n) = \frac{P_{\text{BS}_{\text{serving}}}^{(i)}(n)}{P_{\text{Inter}}^{(i)}(n) + N_0 W_i F_{\text{MS}_i}} \tag{10.18}$$

where N_0 is the received noise spectral density; W_i the sub-carrier bandwidth; and F_{MS_i} is the noise figure at the mobile station. The method followed in equations (10.16)–(10.18) for the derivation of the SINR is perfectly general. It was adapted to OFDM system level simulations in [1, 33], and adopted in the system level simulator platform for the computation of the SINR in each data sub-carrier k, as given by

$$\text{SINR}^{(k)}(n) = P^{(k)}(n) \cdot \overline{G} \cdot \left(\frac{N}{N + N_p}\right) \cdot \frac{R_D}{N_{SD}/N_{ST}} \tag{10.19}$$

where $P^{(k)}(n)$ is the frequency selective fading power profile for the serving cell (propagation from interfering cells is modeled as flat fading); $\overline{G}$ if the Geometric Factor between the mobile station and its serving and interfering cells and is given by equation (10.20); N is the FFT size, including pilot, data and guard sub-carriers; N_p is the cyclic prefix length; R_D is the percentage of maximum total available transmission power allocated to data sub-carriers; N_{SD} is the amount of data sub-carriers per each OFDM symbol; and N_{ST} is the amount of useful (pilot plus data) sub-carriers per OFDM symbol. The Geometric Factor is defined by

$$\overline{G} = \frac{G(\text{Cell}_0, \text{MS}) \times \frac{1}{\text{PL}(\text{Cell}_0,\text{MS}) \times \text{SH}(\text{Cell}_0,\text{MS})}}{\sum_{k=1}^{N} G(\text{Cell}_k, \text{MS}) \times \frac{1}{\text{PL}(\text{Cell}_k,\text{MS}) \times \text{SH}(\text{Cell}_k,\text{MS})} + N_0 W F_{\text{MS}}} \tag{10.20}$$

In equation (10.20) Cell_0 is the serving cell and N is the total amount of interfering base stations.

Assuming that multi-path fading magnitudes and phases, respectively $M_p(n)$ and $\theta_p(n)$, are constant over the frame interval for each path p of the tapped delay channel filter, the frequency-selective fading power profile for the kth sub-carrier is given by [1]

$$P^{(k)}(n) = \left| \sum_{p=1}^{N_{\text{paths}}} M_p A_p \exp\left(j\theta_p\right) \exp\left(-j2\pi f_k T_p\right) \right|^2 \tag{10.21}$$

where p is the tap index (from 1 to 6) of the tapped delay model; A_p is the amplitude value of the long-term average power for the pth tap of the tapped delay filter; T_p is the relative time delay of the pth tap of the tapped delay filter; and f_k is the relative frequency offset of the kth sub-carrier within the spectrum of the OFDM symbol. Parameters A_p and T_p depend on the type of ITU channel used in the modeling of multi-path channel propagation.

Mobile WiMAX standard is an OFDM-based technology. If one designates the set of sub-carriers available for data transmission in each OFDM symbol as N_{data}, the power available at the base station for data transmission (not considering the power boost used in the transmission of pilot sub-carriers, used in channel estimation) by P_{data}, and if one splits this power uniformly over the set of data sub-carriers, the power assigned for the transmission of data sub-carrier n, $n \in [0, \ldots, N_{data} - 1]$ is given by

$$P_{data}^{(n)} = \frac{P_{data}}{N_{data}} \tag{10.22}$$

10.3.10 Performance Metrics

In the execution of each transmission time interval a number of statistics are collected for the computation of the metrics used in the evaluation of the performance of the system level simulation platform. These performance statistics are generated as outputs from the system level simulations and are used in the performance evaluation of the used scenarios and proposed algorithms. The following parameters are used as inputs for the computation of performance metrics:

- Simulation time per run: T_{sim}.
- Number of simulation runs: D.
- Total number of cells being simulated: N_{cells}.
- Total number of users in cells of interest (cells being simulated): N_{users}.
- Number of packet calls for user u: p_u.
- Number of packets in ith packet call of user u: $q_{i,u}$.

Appendix E presents an extensive list with all performance metrics used in all system level simulations performed in the scope of this work.

10.3.11 Traffic Models

In the simulations widespread traffic models have been used for system validation and for characterizing typical mobile applications. The traffic models include:

- Full Queue (FQ) traffic model in which we assume that there is an infinite amount of data bits waiting in the queue of each active user in the system. That is, users are designated as backlogged. This traffic model is particularly interesting in evaluating the maximum capacity of the network.
- Voice over IP traffic model.
- Near Real Time Video with an average source bit rate of 32 kbps, 2 Mbps and 10 Mbps.
- World Wide Web (WWW) traffic model with a source bit rate of 64 kbps, 2 Mbps and 10 Mbps.
- File Transfer Protocol (FTP) traffic model with a source bits rate of 64 kbps and 384 kbps.

Appendix C provides a detailed description of each one of these traffic models.

10.4 Energy Efficient Resource Scheduling

10.4.1 Cross-Layer Framework for Energy Efficient Resource Allocation

Energy-efficient design needs a cross layer approach as all facet of system design create an impact on power consumption ranging from silicon to applications. The authors in [20] particularly focus on a system-based approach towards energy optimal transmission and resource management across time, frequency, and spatial domains. The framework model for EE is developed [34] and shown in Figure 10.8. The power consumption in the transmit mode will be even higher for long-distance communications, such as in cellular networks, and beyond that radio interfaces account for 50% of the overall system energy budget. Cross-layer approaches exploit interactions between different layers and can significantly improve energy-efficiency as well as adaptability to service, traffic, and environment dynamics.

Focusing on the medium access control (MAC) layer, this deals with wireless resources or so namely transport channels, and decides how to allocate the users to the available resources. A MAC that is well designed should be

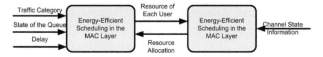

Figure 10.8 Framework of EE based cross-layer resource allocation.

able to maximize the utility of the network whilst ensuring the desired QoS. Beyond that, dynamic resource allocation policies are extended to be also energy-aware since this now plays a pivotal role in the energy consumption of the radio interface. Moreover, according to Miao et al. [20] there two types of access. In distributed access schemes, MAC should be enhanced to cut down the number of wasted transmissions that are corrupted by interference of other users or other antenna elements; while in centralized access schemes, efficient scheduling algorithms should exploit the variations across users to maximize overall energy-efficiency of users in the network. From the Shannon capacity, energy-efficiency can only be obtained at the cost of infinite or huge bandwidth and results in zero or very low spectral efficiency. Transmitting with infinite bandwidth will achieve the highest energy efficiency.

The MAC layer can enhance energy efficiency using the following three measures [20]:

- Energy can be saved in mobile devices by shutting down system components when inactive. The MAC can enable inactive periods by scheduling shutdown intervals according to buffer states, traffic requirements, and channel states.
- The MAC layer controls medium access to assure both individual QoS and network fairness. In distributed access schemes, MAC should be improved to reduce the number of wasted transmissions that are corrupted by interference of other users; while in centralized access schemes, efficient scheduling algorithms should exploit the variations across users, to maximize overall energy efficiency of users in the network.
- Power management at the MAC layer reduces the standby power by developing a tight coordination between users such that they can wake up precisely when they need to transmit or receive data.

10.4.2 Energy Consumption in Classical Resource Scheduling Algorithms

The three main state-of-the-art resource scheduling algorithms are discussed here in order to have an insight into the energy consumption of those scheduling methods and define a type of benchmark. These scheduling schemes are maximum carrier-to-interference (Max C/I), round robin (RR) and proportional fairness (PF). The scheduling methods are briefly described hereunder.

In the MAX C/I method the users are scheduled to use radio resources based on maximum channel gain. This scheme is straight forward in which users are ranked according to their experienced channel gain. In other words, the user with best CQI is ranked up on the top and scheduled to utilize the physical resource blocks for the specific time. The user with the next best CQI condition is then scheduled to utilize physical resource block (PRBs). The ranking M can be found using the following equation $M = \arg \max_m (\beta(t))$ where β is the vector of experienced channel gain by the cell users in time t.

In round robin scheduling, the radio resources are allocated to users in a round-robin fashion. The first selected user is served with whole frequency spectrum for a specific time period and then these resources are revoked back and assigned to the next user for another time period. The previously served user is placed at the end of the waiting queue so that it can be served with radio resources in next round. The new arriving requests are also placed at the tail of the waiting queue. This scheduling continues in the same manner. This scheme offers fairness among the users in radio resource assignment, but it is not practical as one user is served at a time and thus degrading the whole system throughput considerably.

In the PF algorithm, for PRB m, the highest ranked user n' is scheduled to transmit:

$$n' = \arg \max_n \left(\frac{R_{n,m}(t)}{T_n(t)} \right)$$

where $R_{n,m}(t)$ denotes the instantaneous achievable rate at PRB m and $T_n(t)$ is the user's average throughput. Figure 10.9 demonstrates the comparison of the energy consumption of the different classical resource scheduling algorithms. As expected Max. C/I consume least energy compare to other two algorithms due to its channel aware nature. RR is the most energy consuming scheduling policy algorithm since its not channel aware, whereas PF stands in the middle of two since its takes into account both channel information and average data throughput.

10.4.3 Relay Scheduling

Relays are a quite recent architectural concept for infrastructure networks [31]. A relay acts as an intermediate proxy or bridge, by receiving and re-

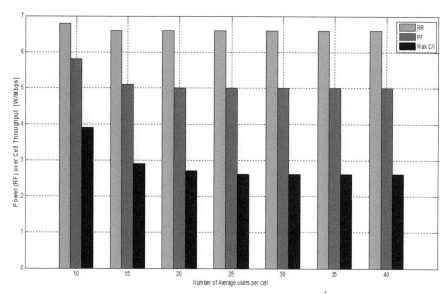

Figure 10.9 Comparison of energy consumption classical resource scheduling algorithms.

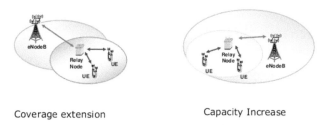

Coverage extension   Capacity Increase

Figure 10.10 Relay deployment approaches.

transmitting a radio packet to extend coverage at the cell edge, or increase capacity. Figure 10.10 shows more flexible and cost-effective deployment options [2], and also being considered in 3GPP LTE.

Relays nodes (RNs) cover much smaller areas than macro cells and thus have significantly lower transmit power compared to the macro eNBs. That is, relay nodes built for small transmission ranges are expected to consume less power and relay nodes can be considered as a promising solution to increase the energy efficiency of a mobile network.

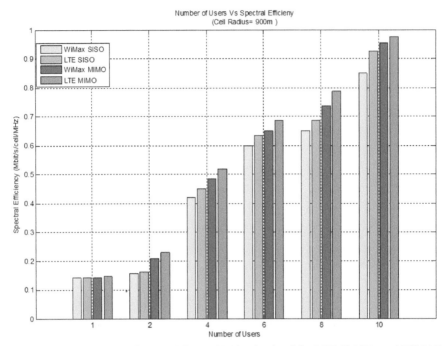

Figure 10.11 Spectral efficiency SISO vs. MIMO (simulated for LTE 10 MHz and WiMAX 5 MHz).

Table 10.7 Modulation and Coding Schemes (MCSs) for simulated LTE, 10 MHz bandwidth.

MCS	Modulation	Coding rate	Maximum Bit Rate (Mbps) (SISO)	Maximum Bit Rate (Mbps) (MIMO)
1	QPSK	1/2	4.32	8.64
2	QPSK	3/4	6.3	12.6
3	16QAM	1/2	16.8	33.6
4	16QAM	3/4	25.2	50.4
5	64QAM	1/2	25.2	50.4
6	64QAM	3/4	38.7	77.4

10.4.4 Energy Analysis SISO vs. MIMO

Energy efficiency of SISO and MIMO were evaluated using the system level approaches given in Section 10.3. The SISO and MIMO performances were analyzed using physical layer standards from both LTE and WiMAX standards. LTE standard already specify the Alamouti-based [7] space-frequency block coding (SFBC) technique for MIMO. A similar approach was made for the WiMAX.

Table 10.8 Modulation and Coding Schemes (MCSs) for simulated WiMAX, 5 MHz bandwidth.

MCS	Modulation	Coding rate	Maximum Bit Rate (Mbps) (SISO)	Maximum Bit Rate (Mbps) (MIMO)
1	QPSK	1/2	2.1	4.2
2	QPSK	3/4	3.2	6.4
3	16QAM	1/2	8.64	17.28
4	16QAM	3/4	12.96	25.92
5	64QAM	1/2	12.96	25.92
6	64QAM	3/4	19.44	38.88

The spectral efficiency results for both the LTE and WiMAX are presented in Figure 10.11. Tables 10.7 and 10.8 shows the maximum bit rate of SISO, and 2×2 SFBC systems employing different MCSs.

The results obtained with the SFBC MIMO in the LTE with 10 MHz bandwidth is the most energy efficient scheme because of diversity gain when the spectral efficiency approaches 1 Mb/s/cell/MHz.

10.5 Alternative Approaches on System Level Simulation

In terms of system simulations several alternative and complementary approaches can be found compared to the approach presented in this chapter. There are works available in the literature regarding the simulation of mobile networks for Beyond 3G and 4G evolutions such as LTE, WiMAX and HS-DPA. Some of these proposals are based on the use of proprietary simulation platforms and others are based on open platforms for system simulation such as Network Simulator 2 (NS2).

Yongquan et al. [30] and Belghith and Nuaymi [4] present the performance of a WiMAX OFDMA system based on system level simulations for multi-user scheduling algorithms in the frequency domain. It is shown that multi-user scheduling in frequency domain can potentially improve OFDMA system efficiency in frequency-selective broadband channels. In [28] the authors propose to estimate the capacity of a WiMAX network using a module developed for NS2 simulation tool. Different scheduling algorithms are considered for the estimation of the spectrum efficiency of a WiMAX network scenario and the results obtained are compared against theoretical values.

In [9, 17] a fixed WiMAX network based on the IEEE 802.16d standard is simulated at the system level. Detailed performance comparisons of two different scenarios with tight frequency reuse schemes are presented.

In [10] detailed link and system level simulations have been performed in interference limited cellular environments for tight 1×1 and 1×3 frequency reuse schemes. The authors conduct exhaustive system level simulations to infer about the dependency of the achieved service throughput, modulation and coding distribution and channel utilization on the applied system load.

In [23] similar work to the one described in this chapter is performed. Namely, the authors elaborate on the implementation of a DRA protocol architecture based on cross-layer signaling in MC-CDMA networks. The proposed DRA is intended to support a very large amount of users with inherent flexibility and granularity necessary to support heterogeneous traffic with limited complexity. Extensive system level simulations are performed to evaluate the DRA performance under such network scenario.

In [13] an extensive set of simulations are performed for verifying the effectiveness of Mobile WiMAX QoS mechanism for different types of traffic profiles, in managing traffic generated by data and multimedia sources. The authors concluded that the performance of the system depends on such factors as the frame duration, the mechanisms used for requesting uplink bandwidth and offered load. Uplink and downlink packet scheduling and transmission are simulated. In [5, 6] dynamic system level simulations are performed to infer whether commonly used schedulers available in the literature are able to guarantee QoS requests of VoIP traffic users, on a scenario of mixed VoIP and WWW traffic users. Algorithms are divided into two different sets: QoS-differentiated and non-QoS-differentiated algorithms.

In [11] a new scheduling approach is proposed based on an strategy where users are temporarily removed from the active set of users considered in the scheduler if the channel quality is lower than a given admission threshold. The temporary removal is easily combined with any conventional scheduling technique providing considerable performance benefits. An exhaustive set of system level simulations are conducted to infer about the benefits of such strategy over simpler scheduling algorithms (without the removal strategy). Srinivasan et al. [32] present the results from extensive simulations to evaluate the downlink performance of Mobile WiMAX employing MIMO channel and Bian and Nix [35] analyze the performance of the same Mobile WiMAX networks according to different PHY layer configurations. Handover and link adaptation are jointly used to limit the amount of inter-cell interference and improve cell coverage and service satisfaction.

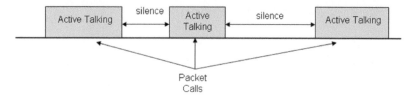

Figure 10.A.1 VoIP traffic model: active talking silence periods.

10.6 Future Work

System level evaluation for energy efficiency optimization is an important engineering task for both manufacturer and operators. In fact, there are still key challenges that need to be addressed in a bid to minimize the energy bill for operators, and to promote a new generation of mobile handsets that are energy conscious. These include:

- Validation of specific 4G scenarios, i.e. LTE-A and also newly emerging technologies related to sensor networks.
- Common Radio Resource Management (CRRM) for different 4G wireless technologies that include efficient cooperative load balancing for energy saving.
- New technologies like Coordinated Multipoint for Energy Efficiency should be analyzed.

10.A Appendix A: Traffic Models

This annex describes the steps followed in the implementation of the traffic models used in the system level simulations performed in this work.

10.A.1 Voice over IP (VoIP) Traffic Model

VoIP refers to real-time delivery of voice packets across networks using the Internet Protocols. A VoIP session is assumed to last for the whole simulation period in each run, since the activation of the user.

A typical phone conversation is marked by periods of active talking/talk spurts (ON periods) interleaved by silence/listening (OFF periods) as illustrated in Figure 10.A.1. VoIP traffic model is modelled by a simple 2-state Markov chain model as shown in Figure 10.A.2 [38].

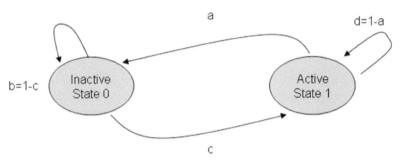

Figure 10.A.2 State transition for VoIP traffic model.

10.A.1.1 Model Statistics

- The conditional probability of transition from active speech state (state 1) to the inactive or silent state (state 0) while is state 1 is a.
- The conditional probability of transition from inactive state 0 to the active state 1 is c.
- The model is updated at the speech encoder frame rate $R = 1/T$, where T is the speech encoder frame duration, assumed as 20 ms.
- The steady-state equilibrium requires that the probabilities of being in state 0 and 1 are given respectively by $P_0 = a/(a + c)$ and $P_1 = c/(a + c)$.
- The voice activity factor (VAF) $\lambda = P_1 = c/(a + c)$
- A talk-spurt is defined as the time period τ_{TS} between entering and leaving the active state. The probability that a talk-spurt lasts during n speech frames is given by $P_{\tau_{TS}=n} P_{\tau_{TS}}(n) = a(1 - a)^{n-1}$, $n = 1, 2, \ldots$.
- The probability that a silence period lasts during n speech frames is given by $P_{\tau_{SP}=n} P_{\tau_{SP}}(n) = c(1 - c)^{n-1}$, $n = 1, 2, \ldots$.
- The mean talk-spurt duration in speech frames is given by $\mu_{TS} = E[\tau_{TS}] = 1/a$.
- The mean silence period in speech frames is given by $\mu_{SP} = E[\tau_{SP}] = 1/c$.
- Since the state transitions from state 1 to state 0 and vice-versa are independent, the mean time between active state entries is given by the sum of the mean time in each state: $\mu_{AE} = \mu_{TS} + \mu_{SP}$ and the mean rate of arrival of transitions into the active state is given by $R_{AE} = 1/\mu_{AE}$.

During the active state, packets of fixed size are generated. The size of the packet and the rate at which the packets are sent depends on the corresponding voice codec and compression schemes. In all simulations conducted in

Table 10.A.1 Detailed description of the VoIP traffic model for IPv4.

Parameter	Values
Codec	RTP AMR 12.2k kbps
	Source Rate 12.2 kbps
Encoder frame length	20ms
Voice Activity Factor (VAF)	40%
Payload	33 bytes
Protocol overhead with compressed header	RTP/UDP/IP: 3 bytes
	IEEE802.16: Generic MAC header: 6 bytes
	CRC for HARQ: 2 bytes
Total voice payload on air interface	44 bytes

this work an AMR codec with a fixed bit rate of 12.2 kbps was assumed. As the encoder frame duration is equal to 20 ms, the packet payload is equal to 244 bits. Although AMR codecs vary the encoding rate from 4.75 to 12.2 kbps, according to the quality of the speech frames reported, link adaptation is not considered in this work.

The length of the VoIP packet must also include the overhead from the TCP/IP layers. Also, the overhead depends on the use of header compression or not. In this work the following configuration is assumed: 6 bytes of MAC header and 2 bytes of HARQ CRC in the IEEE802.16e reference system, 5 bytes for protocol headers (assuming AMR with header compression) and 33 bytes of packet payload. This configuration results in a packet length of 44 bytes. Although the model assumes the generation of packets with silence description as comfort noise, during each inactive state this is not included in the simulations. During each call session the mean duration of an ON state is equal to 1second and the mean duration of an OFF state is equal to 1.5 seconds. Both states follow exponential distributions. During the ON period packets of fixed length are generated at intervals of 20 ms. Table 10.A.1 provides the relevant parameters of the VoIP traffic assumed in the simulations.

For the parameters considered in the VoIP model used in simulations the model statistics are the following: $a = 0.0066$, $c = 0.0044$, $\mu_{TS} = 1/a = 152 \approx 3s$, $\mu_{SP} = 1/c = 227 \approx 4.54s$, $\mu_{AE} = 7.54s$ (1508 frame periods), $R_{AE} = 0.132$ talk spurts per second.

Provided a single reservation is made per user per talk-spurt each user will request resources for transmission of VoIP packets in average every 7.54 seconds or 1508 frame periods.

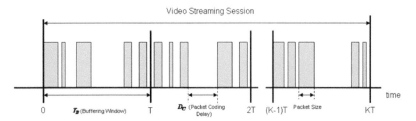

Figure 10.A.3 NRTV packet streaming model.

10.A.2 3GPP near Real Time Video (NRTV) Traffic Model

Figure 10.A.3 describes the steady state of video streaming traffic from the network [37]. Latency at call set-up is not considered in this steady-state model.

A video streaming session is defined as the entire video streaming call time, which is equal to the simulation time for each run in combined snapshot-dynamic mode. Each frame of video arrives at a regular interval T determined by the number of frames per second and each frame is decomposed into a fixed number of slices, each one transmitted within a single packet. The size of each slice is determined according to a Pareto distribution. The encoding delay D introduces delay intervals between the slices of a frame and are modelled by a truncated Pareto distribution. The parameter TB is the length in seconds of the de-jitter buffer window in the mobile station. The de-jitter buffer is used to guarantee a continuous display of video streaming data. It is not used in the generation of data for this model. The video traffic model parameters are defined in Table A.2.

10.A.3 3GPP World Wide Web (WWW) Browsing Traffic Model

The traffic model for WWW bursty traffic used in the simulations is based on the ETSI WWW browsing model [40], but was tailored to reduce simulation run time by decreasing the number of mobile stations required to achieve peak system loading. The main modification is the reduction of the reading time between packet calls. Figure 10.A.4 illustrates the trace of a typical web browsing session. Each packet session consist of multiple packet calls representing web page downloads. Each session is divided into ON and OFF periods. The ON periods correspond to the instants where a web page is downloaded from the server. These are referred as packet calls. The OFF period represent the intermediate reading times required to digest the

Table 10.A.2 NRTV parameters for traffic model.

Information types	Inter-arrival time between the beginning of two consecutive frames	Number of slices in a frame	Slice size	Inter-arrival time between two consecutive slices in a frame	Source bit rate (kbps)
Distribution	Deterministic (10 frames per second)	Deterministic (fixed) N	Truncated Pareto (mean: mu, max: ma)	Truncated Pareto (mean: mu, max: ma)	$N * mu * 8 / T$
Parameters	100 ms	8	K=40bytes; alpha=1.2 mu=50bytes; ma=250bytes	K=2.5ms; alpha=1.2; mu= 6ms max=12.5ms	32 kbps
	//	//	mu=600bytes; ma=3125bytes	//	2 Mbps
	//	//	mu=3125bytes; ma=15625bytes	//	10 Mbps

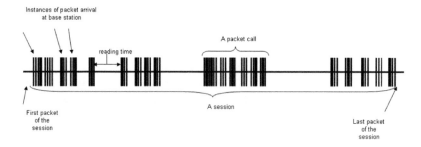

Figure 10.A.4 Web browsing session.

downloaded web page. The size of each packet call (in bytes) is modelled by a truncated Pareto distribution producing a mean packet call size of 25 Kbytes. The reading time is modelled by a geometrically distributed random variable with a mean of 5 seconds. The reading time begins when the mobile station has received the entire packet call. Each packet call is segmented into individual packets. The time interval between two consecutive packets is modelled by a geometric distribution with a mean equal to the ration of the maximum packet size divided by the peak link speed. The size of each packet is fixed and equal to 12000 bits. The "slow-start" TCP/IP rate control mechanism for pacing packet traffic is not implemented.

Table 10.A.3 illustrates the parameters used in this model. Other versions of the same traffic model for higher bit rates are obtained for the parameters in Table 10.A.4.

Table 10.A.3 Web browsing parameters for traffic model.

Process	Distribution	Parameters
Packet Call Size	Truncated Pareto	$\alpha = 1.1$ $k = 4.5$ Kbytes $m = 2$ Mbytes $\mu = 25$ Kbytes
Time between packet calls	Geometric	$\mu t = 5$ seconds
Packet Size	Deterministic	12000 bits
Packet per packet call	Deterministic	Based on packet call size and packet size
Packet inter-arrival time	Geometric	$\mu = $ packet size/peak link speed Peak Link Speed of 2 Mbps

Table 10.A.4 Web browsing parameters for traffic model.

Process	Distribution	Parameters	PBR $= \mu/\mu t$
Packet Call Size	Truncated Pareto	$\mu = 1280$ Kbytes (10,240,000 bits)	2M bps
Packet Call Size	Truncated Pareto	$\mu = 64000$ Kbytes (51,200,000 bits)	10 Mbps

10.A.4 3GPP File Transfer Protocol (FTP) Traffic Model

In FTP applications a session is a sequence of file transfers, separated by reading times. Figure 10.A.5 illustrates a sample packet trace of an FTP session.

The file size in bytes is modelled according to a Log-Normal distribution and the reading time between the end of a download of the previous file and the user request for the next one is modelled according to an exponential distribution with a mean of 180 seconds. The size of each packet is deterministic and equal to 12000 bits.

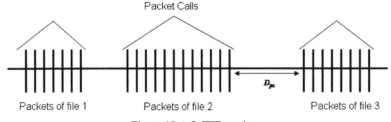

Figure 10.A.5 FTP session.

Table 10.A.5 Web browsing parameters for traffic model.

Process	Distribution	Parameters
File Size	Truncated Log-Normal	Mean = 2 Mbytes
		Standard Deviation = 0.722 Mbytes
		Maximum = 5 Mbytes
		($\mu = 14.45$; $\sigma = 0.35$)
Reading Time	Exponentia	Mean = 180 seconds
		($\lambda = 0.006$)

10.B MIMO Channel Modelling in System Level Simulations

There are different methods for modeling the MIMO channel at the system level. These methods are grouped into two different categories:

- Ray-based: the channel coefficient between each transmit and receive antenna pair is the summation of all rays at each tap of the multi-path filter at each time instant, according to the antenna configuration, gain pattern, angle of arrival (AoA) and angle of departure (AoD) of each ray. The temporal channel variation depends on the travelling speed and direction relative to the AoA/AoD of each ray.
- Correlation based: The MIMO channel coefficients at each tap are mathematically generated according to independent and identically distributed Gaussian random variables, according to the antenna correlation and the temporal correlation, corresponding to a particular Doppler spectrum.

All the simulations conducted along this work were performed by means of the 3GPP Spatial Channel Model for MIMO modeling [36]. In this model the channel gain between each pair of antennas, at both ends of the communication link, results from the superposition of the contributions from each individual path of the tapped delay line model.

10.B.1 3GPP Spatial Channel Model

The Spatial Channel Model (SCM) proposed by 3GPP [36, 39] is a detailed empirical system level model for simulating urban micro-cell, urban macro-cell and suburban macro-cell fading environments. The MIMO channel is represented as a superposition of clustered constituents, with stochastic powers, angles of departure (AoD) and arrival (AoA), as well as times of arrival. Each cluster emulates the effects of small scale fading mechanisms, which contribute to multi-path propagation. The received signal consists of N time-delayed replicas (paths) of the transmitted signal (the number of paths depends on the specific channel model). Each path is associated to a power and delay values, which are randomly generated according to pre-defined random distributions, whose parameters depend on the type of environment used in the simulations. Each path is also associated to clusters of M sub-paths which are not resolvable at the receiver end.

3GPP SCM channel model has a total of $N = 6$ paths and each path has $M = 20$ sub-paths. For the nth path, $\mathbf{H}_n(t)$ denotes the MIMO channel matrix generated by the SCM model. It is a multi-dimensional matrix and is given by

$$\mathbf{H}_n(t) = \begin{bmatrix} h_{11}^n & \cdots & h_{1n_T}^n \\ \vdots & \ddots & \vdots \\ h_{n_R 1}^n & \cdots & h_{n_R n_T}^n \end{bmatrix} \tag{10.B.1}$$

The entry $h_{n_i n_j}^n(t)$ denotes the complex channel gain (complex amplitude) between the n_i receiving antenna and the n_j transmitting antenna for the nth path. It is generated by the superposition of a number of sinusoidal signal components corresponding to each sub-path. The number of rows is equal to the number of receiving antennas, N_R, in the base station linear antenna array and the number of columns is equal to the number of transmitting antennas, N_T, in the mobile station linear antenna array.

The overall procedure for generating the channel matrices consists of three basic steps:

1. Specification of the environment in which the empirical channel model is to be used: suburban macro, urban macro or urban micro.
2. Derivation of the channel parameters to be used in the simulations associated with that environment.
3. Generation of the channel coefficients based on the derived parameters. This step gives the complex channel gains of the channel matrix $\mathbf{H}_n(t)$ for each path.

Once the scenario has been chosen and the location of the base stations (with desired geometry and inter-base distances) has been determined, one may start instantiating users in the area of interest. This entails first randomly generating the user locations and specifying other user-specific quantities such as their velocity vector $\mathbf{v}$, with its direction drawn from a uniform $[0,360°)$ distribution. The specifics of the mobile station antenna or antenna array have to be determined, such as the array orientation, Ω_{MS}, also drawn from a

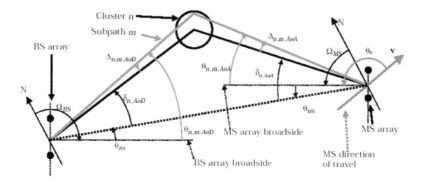

Figure 10.B.1 SCM model for MIMO channel system level simulations.

uniform $[0,360°)$ distribution. The derivation of the channel parameters in step 2 is based on the execution of random distributions whose configuration parameters were previously derived from measurement campaigns, and which depend on the type of environment defined in step 1. Figure 10.B.1 depicts the 3GPP SCM channel model for MIMO simulations where only one cluster of scatterers is shown. Two uniform linear arrays at both ends of the communication path, with two antennas each, are illustrated in this figure.

The following parameters are derived directly from the cell layout, mobile positions and input configurations:

- θ_{BS} is the angle between BS-MS line-of-sight and the BS broadside array.
- θ_{MS} is the angle between BS-MS line-of-sight and the MS broadside array.
- d_{n_T} is the distance, in meters, from base station antenna element n_T from the reference $n_T = 1$ antenna (for the reference antenna $n_T = 1$, $d_1 = 0$).
- d_{n_R} is the distance in meters from mobile station antenna element n_R from the reference $n_R = 1$ antenna (for the reference antenna $n_R = 1$, $d_1 = 0$).

The remaining parameters such as the angle of arrival (AoA) and departure (AoD) of each ray are generated according to specific distributions and parametrization which depend on the type of environment and channel model being simulated. These parameters are assumed as constant over the whole simulation (in case of mobile stationarity) or as changing in a very slowly time scale:

- $\theta_{n.m,\text{AoD}}$ is the angle of departure (AoD) for the mth sub-path of the nth path at the base station with respect to the base station broadside. It is computed as follows: $\theta_{n.m,\text{AoD}} = \theta_{\text{BS}} + \delta_{n,\text{AoD}} + \Delta_{n,m,\text{AoD}}$. The parameters $\delta_{n,\text{AoD}}$ and $\Delta_{n,m,\text{AoD}}$ denote, respectively, the AoD and the offset for the mth sub-path of the nth path with respect to $\delta_{n,\text{AoD}}$.
- $\theta_{n.m,\text{AoA}}$ is the angle of arrival (AoA) for the same mth sub-path of the same nth path with respect to the mobile station broadside. It is computed as follows: $\theta_{n.m,\text{AoA}} = \theta_{\text{MS}} + \delta_{n,\text{AoA}} + \Delta_{n,m,\text{AoA}}$. The parameters $\Delta_{n,m,\text{AoA}}$ and $\delta_{n,\text{AoA}}$ denote, respectively, the AoA and the offset for the for the mth sub-path nth path with respect to $\delta_{n,\text{AoA}}$.

These parameters are used in the computation of the complex channel $h^n_{n_R n_T}(t)$ between the receiving antenna n_R and transmitting antenna n_T, according to

$$h^n_{n_R,n_T}(t) = \sqrt{\frac{P_n \sigma_{SF}}{M}} \sum_{m=1}^{M} \tag{10.B.2}$$

$$\times \begin{pmatrix} \sqrt{G_{BS}(\theta_{n.m,AoD})} \exp(j\left[kd_{n_T}\sin(\theta_{n,m,AoD}) + \Phi_{n,m}\right])x \\ \sqrt{G_{MS}(\theta_{n,m,AoA})} \exp(jkd_{n_R}\sin(\theta_{n,m,AoA}))x \\ \exp(jk\,\|\mathbf{v}\|\cos(\theta_{n,m,AoA}) - \theta_{\mathbf{v}})t) \end{pmatrix}$$

- P_n is the power associated to the nth path, which is modeled as an exponential Power Delay Profile (PDP) by the model.
- σ_{SF} is the log-normal shadow fading applied to the N paths for a given mobile drop.
- $G_{\text{BS}}(\theta_{n.m,\text{AoD}})$ and $G_{\text{MA}}(\theta_{n.m,\text{AoA}})$ are, respectively, the base station and mobile station antenna gains of each array element.
- $k = 2\pi/\lambda$ is the wave number.
- $\|\mathbf{v}\|$ is the magnitude and θ_v is the angle of the mobile station velocity vector.

For a given scenario, realizations of each user's parameters such as: the path delays, powers and sub-paths, angles of departure and arrival can be derived using the procedures described in detail in [39].

The following steps are required to generate the channel element of one sub-path of a multipath from the link between a desired user receive antenna and a particular transmit antenna at the base station:

1. Choose an environment.
2. Determine distances and orientation parameters.

3. Determine the Delay Spread (DS), Angle Spread (AS) and Shadow Fading (SF) components. DS and SF are correlated and log-normally distributed. AS is also log-normally distributed and correlated to DS and SF.
4. Determine random delays $(\tau_n, n = 1, \ldots, N)$ for each of the N multipath components.
5. Determine random average powers $(P_n, n = 1, \ldots, N)$ for each of the N multipath components.
6. Determine AoDs for each one of the N multipath components at the base station.
7. Associate the multipath delays with AoDs.
8. Determine the powers, phases and offsets AoDs of the $M = 20$ sub-paths for each one of the N multi-paths at the base station.
9. Determine AoAs for each one of the N multipath components at the mobile station.
10. Determine the offset AoAs at the mobile station for each one of the $M = 20$ sub-paths of each one of the N paths.
11. Associate the base station and mobile station paths and sub-paths.
12. Determine the antenna gains of the base station and mobile station sub-paths as a function of their respective sub-path AoDs and AoAs.
13. Apply the path loss based on the base-station to mobile station distance from Step 2 and the log-normal shadow fading determined in Step 3 as large-scale parameters to each one of the sub-path powers of the channel model.
14. Generate the channel coefficients.

The SINR at each receive antenna is computed for each path of the model and they are all combined at the SINR level (using the EESM mapping method). In order to do so interference from all other interferers is explicitly modeled in system-level simulations. It is assumed that due to OFDM cyclic prefix transmitted in each symbol intra-cell interference is neglected, i.e., only inter-cell interference is assumed in the simulations.

Figure 10.B.2 illustrates the scenario assumed (cellular and base station layout, as well as mobile stations and antenna array orientation) for conducting MIMO channel simulations in the system level simulator. In particular, it differs from Figure 10.B.1 (reprinted from 3GPP specifications), as it describes the steps followed in the computation of the values assumed for the different parameters, which are used in the MIMO channel model for conducting system level simulations, according to 3GPP specifications. After

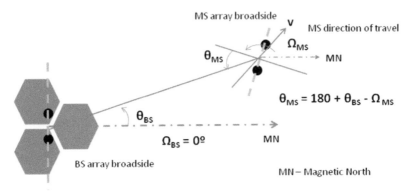

Figure 10.B.2 Network layout for MIMO channel computation.

deploying the cells and base stations in the network layout, the angles θ_{BS} and Ω_{BS} result automatically defined. The angles θ_{MS} and Ω_{MS}, regarding each mobile station dropped in the network, depend on the orientation of each mobile station antenna array and on the direction followed by each mobile moving along the coverage area pertaining to its serving base station. These parameters are computed and modified dynamically, as the simulation evolves.

As mentioned before, these angles are central to the computation of the complex gains for each pair of transmitting and receiving antennas in the MIMO channel matrix.

10.B.2 Modeling the SINR at the Mobile Station

In what follows the downlink communication is assumed (although the same explanation is also valid for the uplink communication). Also, the channel can be assumed as symmetrical because all system level simulations are performed for the time division duplex mode (TDD). As it was mentioned in previous sections, for the computation of the performance metrics in system level evaluations, the figure of merit used is the BLER, which is obtained from the look-up tables, according to the SINR value computed at the mobile station. The derivation of SINR for the MIMO channel is performed in two separate parts:

- Computation of the desired user radio signal arriving from the mobile station's serving cell.
- Computation of the interfering user radio signal arriving from neighboring cells.

10.B.2.1 Modelling the Desired User Signal at the Mobile Station

The desired signal from the serving cell is computed according to what is described in the previous section. After the derivation of the channel matrix for each one of the N paths the channel response for each pair of transmitting-receiving antenna is derived from equation (10.B.2) by combining the contributions from all paths.

10.B.2.2 Modelling Inter-Cell Signal Interference at the Mobile Station

Although the spatial characteristics of the signals received from the serving as well as from interfering cells can be modeled according to the empirical MIMO channel model, it is very complex and computationally intensive to explicitly model the MIMO channel from all interfering cells, especially those cells from which receiving powers are relatively weak. The performance difference achieved when modeling signals from relatively weak interferers as spatially white (ignoring their spatial characteristics) is negligible.

In a MIMO channel the modeling of other-cell interference is done by considering three types of neighboring cells: near strong cells whose interference is modeled by a MIMO channel, near weak cells whose interference is modeled as wideband (frequency-selective) channel and far sectors whose interference is modeled as narrowband (flat) channel. According to this approach, and to simplify system performance evaluation, the set of neighboring cells can be divided into three groups: MIMO (list-A), SISO wideband list (list-B) and SISO narrowband list (list-C). These lists are filled according to the values of path loss and shadowing to each cell, i.e., the cells are ranked in order of the received power. Strongest A cells are inserted into list A whose size is a trade-off between computational complexity and performance. 3GPP recommends a number of 8 cells for list-A in a 3-sectored cell deployment [39]. The next B cells below list A in the rank are inserted as members of list B and the remaining ones are inserted as members of list C.

Interference from Strong Interferers

The strong interferers are modeled according to the 3GPP SCM MIMO channel model. The interference from one interferer over MIMO coming to each receive antenna is collected from all multi-paths ($N_T x N$). Then the total interference for each receiving antenna is computed as the sum of the interference from all interferers in the list-A, according to

$$I_{\text{list}-A} = \sum_{a \in \text{list}-A} \sum_{i=1}^{N_T} \sum_{j=1}^{N} P_{a,i,j} \tag{10.B.3}$$

where $p_{a,i,j}$ is the received power over time from the ith transmit antenna for the jth, $j = 1, \ldots, N$ path of the wideband channel model of the ath cell in list-A.

Interference from Weak Interferers

The weak interferers are modeled as spatially while Gaussian noise processes whose variances are based on a multi-path Rayleigh fading process (wideband SISO channel), depending on the simulation environment. The fading processes for each cell and receive antenna are independent and equivalent for each mobile receive antenna. The total received noise power from cells in list-B, at the n_Rth antenna, is given by

$$I_{\text{list}-B} = \sum_{b \in \text{list}-B} \sum_{i=1}^{N} p_{b,n_R,i} \tag{10.B.4}$$

where $p_{b,n_R,i}$ is the received power over time for the n_Rth receive antenna, coming from the ith, $i = 1, \ldots, N$ path of the wideband channel model of the bth cell in list-B.

Interference from List-C Interferers

Interferers in list-C are modeled as spatially white Gaussian noise processes whose variances are based on a flat Rayleigh fading process (single path). The fading processes for each cell and receive antenna are independent and the fading is equivalent for each mobile receive antenna. The total received noise power at the n_Rth receive antenna, due to all cells in list-C is given by

$$I_{\text{list}-C} = \sum_{c \in \text{list}-B} p_{c,n_R} \tag{10.B.5}$$

where p_{c,n_R} is the received power over time for the n_Rth receive antenna, coming from the cth cell in list C.

The total amount of inter-cell interference on the n_Rth antenna is given by the contribution from cells in the three lists, according to

$$I_{\text{total}} = I_{\text{list}-A} + I_{\text{list}-B} + I_{\text{list}-B} \tag{10.B.6}$$

10.B.3 Link to System Interface for MIMO Channel

The implementation of a MIMO scheme in the radio access of a wireless system is intended to increase transmission diversity in order to improve the BER, increase the transmitted data rate or trade-off both. Transmit diversity is achieved by means of Space Time Block Coding schemes (STBC) of which the Alamouti's STBC [7], with two antennas at the base station and one antenna at the mobile station, is an example. The 2×2 Alamouti's STBC scheme turns out to be a receiver simplification because the coding is performed at the transmitter. In the 2×2 Alamouti's STBC scheme a Maximum Ratio Combiner (MRC) receiver increases the diversity of the signal at the mobile station receiver. Vertical 2×2 BLAST (VBLAST) [4, 9, 28, 30] is an example of a spatial multiplexing scheme where two independent symbols are transmitted at each antenna in the base station and recovered at the mobile's antenna array with a Maximum Likelihood Decoder (MLD), Minimum Mean Square Error (MMSE) or Zero Forcing (ZF) receiver. The modeling of the transmission chain of the link layer depends highly on the type of MIMO scheme implemented as well as on the performance results obtained.

The WiMAX Forum has selected two different multiple antenna profiles for use on the downlink. In Mobile WiMAX the first multiple antenna profile is the simple 2×1 or 2×2 Alamouti STBC scheme referred to as Matrix A in the specifications. In OFDMA-based WiMAX system this technique is applied sub-carrier by sub-carrier. The second multiple antenna profile included is the 2×2 SM scheme referred to as Matrix B in the specifications. Only Matrix A MIMO was implemented in the simulator.

Matrix A MIMO Implementation

Matrix A is denoted as $\begin{bmatrix} ccs_1 & s_2^* \\ s_2 & -s_1^* \end{bmatrix}$. The rows represent the two antennas at the transmitter and the columns represent two adjacent transmission OFDM symbols. During the first transmission period transmit antenna 1 transmits symbol s_1 and transmit antenna 2 transmits symbol. During the second transmit period transmit antenna 1 transmits symbol and transmit antenna 2 transmits symbol s_2. The optimum receiver estimates the transmitted symbols s_2^* and $-s_1^*$ as illustrated in the following equation:

$$x_1 = \left(|h_{11}|^2 + |h_{12}|^2 + |h_{21}|^2 + |h_{22}|^2 \right)^2 s_1 + n_1$$

$$x_2 = \left(|h_{11}|^2 + |h_{12}|^2 + |h_{21}|^2 + |h_{22}|^2 \right)^2 s_2 + n_1 \qquad (10.B.7)$$

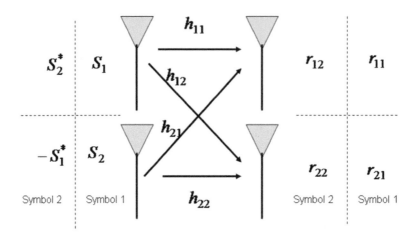

Figure 10.B.3 STBC Alamouti scheme.

In equation (10.B.7) s_1 and s_2 correspond to the signal at receiving antenna 1 and 2, respectively. These expressions clearly show that the impact of the Alamouti scheme is an enhancement of the channel conditions by a fourth diversity order. This corresponds to improvements in the BER. It is also clear that the two signals can be completely separated at the receiver as they do not interfere to each other. The SINR for the information transmitted along two consecutive symbols and sub-carrier k is given by

$$SINR_k = \overline{G}\left(|h_{11}^k|^2 + |h_{12}^k|^2 + |h_{21}^k|^2 + |h_{22}^k|^2\right) \qquad (10.B.8)$$

where $\overline{G}$ is the user's geometric factor. This value is used in the computation of the compressed scalar SINR value for the derivation of the BLER.

10.C Performance Metrics for System Level Simulation

In this appendix the performance statistics, generated as an output from the system level simulations, and used in the performance evaluation of the used scenarios and proposed algorithms are described. Metrics are collected along a simulation run.

For a simulation run:

- Simulation time per run: T_{sim}.
- Number of simulation runs: D.
- Total number of cells being simulated: N_{cells}.

- Total number of users in cells of interest (cells being simulated): N_{users}.
- Number of packet calls for user u: p_u.
- Number of packets in the ith packet call of user u: $q_{i,u}$.

10.C.1 Throughput Performance Metrics

10.C.1.1 Average Service Throughput per-Cell

The average service throughput per cell is defined as the sum of the total amount of bits successfully received by all active users in the system, divided by the product of the number of cells simulated and the simulation duration:

$$R_{\text{service}}^{\text{DL(UL)}} = \frac{\sum_{u=1}^{N_k^{\text{users DL(UL)}}} \sum_{i=1}^{p_{u,k}^{\text{DL(UL)}}} \sum_{j=1}^{q_{i,u,k}^{\text{DL(UL)}}} b_{j,i,u}}{N_{\text{cells}} T_{\text{sim}}} \tag{10.C.1}$$

where $N_k^{\text{users,DL(UL)}}$ is the number of users transmitting in DL(UL) in the kth cell; $p_{u,k}^{\text{DL(UL)}}$ is the number of packet calls for user u in cell k; $q_{i,u,k}^{DL(UL)}$ is the number of packets for the ith packet call for user u in cell k; and $b_{j,i,u}$ is the number of bits received with success in the jth packet of packet call i for user u in cell k.

10.C.1.2 Average Over-the-Air (OTA) Cell Throughput (kbps/cell) (3GPP Definition)

The average OTA throughput per cell is defined as the sum of the total amount of bits being successfully received by all active users in the system divided by the product of the number of cells being simulated in the system and the total amount of time spent in the transmission of these packets.

$$R_{\text{OTA}}^{\text{DL(UL)}} = \frac{\sum_{u=1}^{N_k^{\text{users,DL(UL)}}} \sum_{i=1}^{p_{u,k}^{\text{DL(UL)}}} \sum_{j=1}^{q_{i,u,k}^{\text{DL(UL)}}} b_{j,i,u}}{N_{\text{cells}} T_{\text{trans}}} \tag{10.C.2}$$

where T_{trans} is the time required to transmit these packets.

10.C.1.3 Average Over-the-Air (OTA) Cell Throughput (kbps/cell) (Peak Bit Rate Definition)

This metric is very similar to the OTA throughput. But here all bits (correct and erroneous) are considered in its computation.

$$R_{\text{OTA\_PBR}}^{\text{DL(UL)}} = \frac{\sum_{u=1}^{N_k^{\text{users,DL(UL)}}} \sum_{i=1}^{p_{u,k}^{\text{DL(UL)}}} \sum_{j=1}^{q_{i,u,k}^{\text{DL(UL)}}} b_{j,i,u}^{\text{trans}}}{N_{\text{cells}} T_{\text{trans}}} \tag{10.C.3}$$

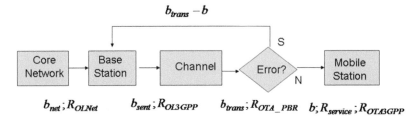

Figure 10.C.1 Traffic metrics interdependence.

where $b_{j,i,u}^{\text{trans}}$ is the number of bits received (with error or success) in the jth packet of packet call i for user u in cell k.

10.C.1.4 Offered Cell Load (kbps/cell) (3GPP Definition)
This metric is used in the evaluation of the data load (in kbps) withdrawn from the base station's buffers for transmission, i.e., the influence of the channel in the transmission of the data is not being considered.

$$R_{\text{OL3GPP}}^{\text{DL(UL)}} = \frac{b_{\text{Sent}}^{\text{DL(UL)}}}{N_{\text{Cells}} T_{\text{sim}}} \tag{10.C.4}$$

where $b_{\text{Sent}}^{\text{DL(UL)}}$ is the total amount of data bits that have been withdrawn from the base station's queues and sent over the air interface for DL(UL) connection, for all mobile stations being simulated over the whole simulation run.

10.C.1.5 Offered Cell Load (kbps/cell) (Network Definition)
This metric measures the offered load from the core network to the base station for all mobile stations being simulated in the system over the whole simulation run.

$$R_{\text{OLNetwork}}^{\text{DL(UL)}} = \frac{b_{\text{Networknet}}^{\text{DL(UL)}}}{N_{\text{Cells}} T_{\text{sim}}} \tag{10.C.5}$$

where $b_{\text{Network}}^{\text{DL(UL)}}$ is the total amount of data bits that have arrived to the base station's queues from the core network over the whole simulation run. Figure 10.C.1 illustrates the relation among these different metrics and figures.

10.C.1.6 User Average Peak Bit Rate at a Given Distance (kbps)
This metric gives the average peak bit rate of a given user at a given distance, d, in steps of 10 m, from the base station. For one user i the average peak bit

rate is defined by

$$R_{\text{PBR}}(i) = \frac{\sum_{k=1}^{N(i)} R_k(i)}{N(i) \cdot T} \tag{10.C.6}$$

where $N(i)$ is the total amount of frames received by the mobile station i (both transmitted and retransmitted frames are taken into account) and $R_k(i)$ is the total amount of bits in the kth frame received by mobile station i.

10.C.1.7 Per User Service Data Throughput

The user's service data throughput is defined as the ratio of the number of information bits successfully received by the user and the total simulation run time. If user u has $p_u^{\text{DL(UL)}}$ downlink (uplink) packet calls with $q_{i,u}^{\text{DL(UL)}}$ packets for the ith downlink (uplink) packet call and $b_{j,i,u}$ bits in the jth packet the average user throughput for user u is given by

$$R_u^{\text{DL(UL)}} = \frac{\sum_{i=1}^{p_u^{\text{DL(UL)}}} \sum_{j=1}^{q_{i,i}^{\text{DL(UL)}}} b_{j,i,u}}{T_{\text{sim}}} \tag{10.C.7}$$

10.C.1.8 Per-User Average Service Throughput

The average per-user service throughput is defined as the sum of the user service throughput of each user divided by the total number of users in the system:

$$\overline{R_u^{\text{DL(UL)}}} = \frac{\sum_{u=1}^{N_{\text{users}}} R_u^{DL(UL)}}{N_{\text{users}}} \tag{10.C.8}$$

10.C.1.9 Average Packet Call Throughput for a User

If there are N_{users} in the cell of interest and $R_{k,u}^{\text{DL(UL)}}$ is the service throughput for the nth user in the cell, the DL or UL service throughput for the cell is given by

$$R^{\text{DL(UL)}} = \sum_{u=1}^{N_{\text{users}}} R_u^{\text{DL(UL)}} \tag{10.C.9}$$

The packet call throughput is equal to the total amount of bits per packet call received with success divided by the duration of the packet call. If user u has $p_u^{\text{DL(UL)}}$ downlink (uplink) packet calls with $q_{i,u}^{\text{DL(UL)}}$ packets for the ith downlink (uplink) packet call and jth packet call then the average packet call throughput is given by

$$R_u^{pc^{\text{DL(UL)}}} = \frac{1}{p_u^{\text{DL(UL)}}} \left(\sum_{i=1}^{p_u^{\text{DL(UL)}}} \frac{\sum_{j=1}^{q_{i,u}^{\text{DL(UL)}}} b_{j,i,u}}{\left(T_{i,u}^{\text{end,DL(UL)}} - T_{i,u}^{\text{start,DL(UL)}} \right)} \right) \qquad (10.\text{C}.10)$$

where $T_{i,u}^{\text{start,DL(UL)}}$ is the time instant at which the transmission of the first packet of the ith DL(UL) packet call for user u starts and $T_{i,u}^{\text{end,DL(UL)}}$ defines the time instant at which the last packet of the ith DL(UL) packet call for user u is received with success. For uncompleted packet calls this parameter is set to the simulation end time.

10.C.1.10 Average Per-User Packet Call Throughput
The average per-user packet call throughput is defined as the sum of the average packet call throughput of each user divided by the total number of users in the system.

$$\overline{R_u^{\text{pc,DL(UL)}}} = \frac{\sum_{u=1}^{N_{\text{users}}} R_u^{pc,DL(UL)}}{N_{\text{users}}} \qquad (10.\text{C}.11)$$

10.C.1.11 Throughput Outage
The throughput outage is defined as the percentage of users with service data rate $R_u^{DL(UL)}$ less than a pre-defined minimum rate $R_{\min}$.

Cell Edge User Throughput
The cell edge user throughput is defined as the 5th percentile point of the CDF of user's average packet call throughput.

10.C.2 Performance Metrics for Delay Sensitive Applications

10.C.2.1 Packet Delay
For an individual packet the delay is defined as the time elapsed between the instant when the packet enters the queue at transmitter and the time when the packet is received successfully by the mobile station. If a packet is not successfully delivered by the end of a run its ending time is the end of the run. Assuming the jth packet of the ith packet call destined for user u arrives at the base station (mobile station) at time $T_{j,i,u}^{\text{arr,DL(UL)}}$ and is delivered with success to the mobile station (base station) at time $T_{j,i,u}^{\text{dep,DL(UL)}}$, the packet

delay is defined as

$$\text{Delay}_{j,i,u}^{\text{DL(UL)}} = T_{j,i,u}^{\text{dep,DL(UL)}} T_{j,i,u}^{\text{arr,DL(UL)}} \tag{10.C.12}$$

10.C.2.2 Use Average Packet Delay
The average packet delay is defined as the average interval between packets
originated at the source station (mobile or base station) and received at the
destination station (base or mobile station) in a system for a given packet call
duration. The average packet delay for user u is given by

$$D_u^{\text{avg,DL(UL)}} = \frac{\sum_{i=1}^{p_u} \sum_{j=1}^{q_{i,u}} \left(T_{j,i,u}^{\text{dep,DL(UL)}} - T_{j,i,u}^{\text{arr,DL(UL)}} \right)}{\sum_{i=1}^{p_u} q_{i,u}} \tag{10.C.13}$$

10.C.2.3 Residual Frame Erasure Rate (FER)
This metric is computed for each user and for each packet service session. A
packet service session contains one or several packet calls depending on the
application. A packet service session starts when the first packet of the first
packet call of a given service begins and ends when the last packet of the last
packet call of the same service has been transmitted. One packet call contains
one or several packets. The Residual FER is given by

$$\text{FER}_{\text{residual}} = \frac{\eta_{\text{dropped\_packets}}}{\eta_{\text{packets}}} \tag{10.C.14}$$

where $\eta_{\text{dropped\_packets}}$ is the total amount of dropped packets in the packet service session and η_{packets} is the total amount of packets in the packet session.
A dropped packet is the one in which the maximum number of transmission
attempts has been achieved without the packet being successfully decoded.

10.C.2.4 Packet Loss Ratio
The packet loss ratio is computed for each user and for each packet service
session and is defined as

$$\text{PDR} = \frac{\eta_{\text{discarted\_packets}}}{\eta_{\text{packets}}} \tag{10.C.15}$$

where $\eta_{\text{discarted\_packets}}$ is the total amount of packets discarded due to time-
out (delay bound violation and maximum number of transmission attempts
achieved).

10.C.2.5 Spectral Efficiency (bps/Hz)

This is the ratio of correctly transmitted bits over the radio resources to the total amount of available bandwidth. The average cell spectral efficiency is defined as

$$SE = \frac{R}{BW_{eff}} \tag{10.C.16}$$

where R is the aggregate cell throughput, BW_{eff} is the effective channel bandwidth, defined as $BW_{eff} = BW * TR$, where BW is the used channel bandwidth and TR is the time ratio of the link. For example for TDD with DL : UL = 2 : 1, TR = 2/3 for DL and 1/3 for UL.

10.C.2.6 System Outage

A user is said to be in outage if more than a given percentage of packets experience a delay greater than a certain time. The system is said to be in outage if any individual users are in outage.

10.C.2.7 System Capacity

System capacity is defined as the maximum numbers of users that can be serviced without making the system exceed the maximum allowed outage probability.

Acknowledgements

The authors would like to acknowledge the project No. 23205 – GREEN-T, co-financed by the European Funds for Regional Development (FEDER) by COMPETE – Programa Operacional do Centro (PO Centro) of QREN. Valdemar Monteiro is a PhD student at Kingston University.

References

[1] 3GPP, Feasibility study for OFDM for UTRAN enhancement. Technical report, 3GPP, TR, 25.892, V1.1.0., 2004.
[2] 3GPP, Evolved universal terrestrial radio access (E-UTRA); further advancements for E-UTRA physical layer aspects (release 9). Technical Report, 3GPP, TR, 36.814,V9.0.0, March 2010.
[3] Nortel Networks 3GPP, Update of OFDM SI simulation methodology. Technical report, 3GPP, R1-030224, February 2003.
[4] A. Belghith and L. Nuaymi, WiMAX capacity estimations and simulation results, in *Proceedings of IEEE Vehicular Technology Conference (VTC 2008)*, May 2008.

[5] A. Braga, E. B. Rodrigues, and F. R. P. Cavalcanti, Novel scheduling algorithms aiming for QoS guarantees for VoIP over HSDPA. In *Proceedings of IEEE VI International Telecommunications Symposium (ITS2006)*, Fortaleza, Brazil, pp. 94–99, September 2006.

[6] A. Braga, E. B. Rodrigues, and F. R. P. Cavalcanti. Packet scheduling for voice over IP over HSDPA in mixed traffic scenarios with different end-to-end delay budgets, in *Proceedings of IEEE VI International Telecommunications Symposium (ITS2006)*, Fortaleza, Brazil, September 2006.

[7] S. M. Alamouti, A simple transmit diversity technique for wireless communications, *IEEE Journal on Selected Areas in Communications*, vol. 16, pp. 1451–1458, 1998.

[8] B. Badic et al., Energy efficient radio access architectures for green radio: Large versus small cell size deployment, in *Proceedings of IEEE 70th Vehicular Technology Conference Fall (VTC2009-Fall)*, September 2009.

[9] C. F. Ball, E. Humburg, K. Ivanov, and F. Treml, Comparison of IEEE802.16 WiMAX scenarios with fixed and mobile subscribers in tight reuse. [Online] Available: `http://www.eurasip.org/Proceedings/Ext/IST05/papers/147.pdf`.

[10] C. F. Ball, E. Humburg, K. Ivanov, and F. Treml, Performance analysis of IEEE802.16e based cellular man with OFDM-256 in mobile scenarios, in *Proceedings of IEEE Vehicular Technology Conference (VTC 2005 Spring)*, pp. 2061–2066, May/June 2005.

[11] C. F. Ball, F. Treml, X. Gaube, and A. Klein, Performance analysis of temporary removal scheduling applied to mobile wimax scenarios in tight frequency reuse, in *Proceedings of IEEE International Symposium on Personal, Indoor and Mobile Radio Communications (PIMRC 2005)*, vol. 2, pp. 888–894, September 2005.

[12] Xiadong Cai, B. Georgios, and G. Giannakis, A two-dimensional channel simulation model for shadowing processes, *IEEE Transactions on Vehicular Technology*, vol. 52, no. 6, pp. 498–523, November 2003.

[13] C. Cicconetti, A. Erta, L. Lenzini, and E. Mingozzi, Performance evaluation of the IEEE 802.16 mac for QoS support, *IEEE Transactions on Mobile Computing*, vol. 6, no. 1, pp. 26–37, January 2007.

[14] C. Tao et al. Energy efficiency metrics for green wireless communications, in *Proceedings of International Conference on Wireless Communications and Signal Processing (WCSP 2010)*, pp. 1–6, 2010.

[15] Universal Mobile Telecommunications System (UMTS) ETSI. Selection procedures for the choice of the radio transmission technologies of the UMTS. Technical Report (UMTS 30.03 version 3.2.0), TR 101 112 v3.2.0, April 1998.

[16] F. Richter et al., Energy efficiency aspects of base station deployment strategies for cellular networks, in *Proceedings of IEEE 70th Vehicular Technology Conference Fall (VTC2009-Fall)*, September 2009.

[17] A. Fernekess, A. Klein, B. Wegmann, K. Dietrich, and E. Humburg, Performance of IEEE 802.16e OFDMA in tight reuse scenarios, in *Proceedings of IEEE International Symposium on Personal, Indoor and Mobile Radio Communications (PIMRC 2007)*, September 2007.

[18] WiMAX Forum, Mobile WiMAX – Part I: A Technical Overview and Performance Evaluation.

[19] WiMAX Forum, WiMAX System Evaluation Methodology V2.1, July 2008.

[20] G. Miao et al., Cross-layer optimization for energy-efficient wireless communications: A survey, *Wirel. Commun. Mob. Comput*, vol. 9, pp. 529–542, 2009.

[21] IEEE 802.16 Broadband Wireless Access Working Group, Draft IEEE 802.16m Evaluation Methodology, October 2007.

[22] M. Gudmundson, Correlation model for shadow fading in mobile radio systems, *Electronics Letters*, vol. 27, no. 4, pp. 2145–2146, November 1991.

[23] D. T. P. Huy, J. Rodriguez, A. Gameiro, and R. Tafazolli, Dynamic resource allocation for beyond 3G cellular networks, *Journal on Wireless Personal Communications*, vol. 43, pp. 1727–1740, April 2007.

[24] ITU, Guidelines for evaluation of radio transmission technologies for IMT-2000. Technical Report, Recommendations ITU-R, M.1225, 1997.

[25] J. Medbo, H. Anderson, P. Schramm, and H. Asplund, Channel models for hiperlan/2 in different indoor scenarios. Technical report, COST259 TD(98), Bradford, UK, April 1998.

[26] W. C. Jakes, *Microwave Mobile Communications*, Wiley, New York, 1974.

[27] Yunxin Li and Xiaojing Huang, The generation of independent Rayleigh faders, in *Proceedings of the IEEE International Conference on Communications (ICC 2000)*, vol. 1, pp. 18–22, June 2000.

[28] A. A. Maltsev and A. V. Pudeyev, H multi-user frequency domain scheduling for WiMAX OFDMA, in *Proceedings of 16th IST Mobile and Wireless Communications Summit*, July 2007.

[29] O. Arnold et al. Power consumption modeling of different base station types in heterogeneous cellular networks, in *Proceedings of Future Network and Mobile Summit*, June 2010.

[30] Yongquan Qiang, G. Vivier, Jing Yang, and Ning Xu, Inter-cell interference modeling for OFDMA systems with beamforming. In *Proceedings of IEEE Vehicular Technology Conference (VTC 2008-Fall)*, September 2008.

[31] EARTH (Energy Aware Radio and neTwork Technologies) FP7 Project INFSO-ICT-247733 EARTH, D3.1 Most Promising Tracks of Green Network Technologies, Deliverable, 2010.

[32] R. Srinivasan, S. Timiri, A. Davydov, and A. Papathanassiou, Downlink spectral efficiency of mobile WiMAX, in *Proceedings of IEEE Vehicular Technology Conference (VTC 2007-Spring)*, pp. 2786–2790, April 2007.

[33] F. Wang, A. Ghosh, R. Love, K. Stewart, R. Ratasuk, R. Bachu, Y. Sun, and Q. Zhao, IEEE 802.16e system performance: analysis and simulations, in *Proceedings of IEEE International Symposium on Personal, Indoor and Mobile Radio Communications (PIMRC 2005)*, vol. 2, pp. 900–904, 2005.

[34] X.-H. Li et al., Resource management algorithm based on cross-layer design for OFDM systems, in *Proceedings of the Seventh International Conference on Parallel and Distributed Computing, Applications and Technologies*, 2006.

[35] Y. Q. Bian and A. R. Nix, WiMAX: Multi-cell network evaluation and capacity optimization, in *Proceedings of IEEE Vehicular Technology Conference (VTC 2008-Spring)*, pp. 1276–1280, May 2008.

[36] 3GPP. Spatial channel model for multiple-input multiple-output simulations. Technical report, 3GPP TR 25.996, May 2003.

[37] 3GPP. Feasibility study for OFDM for UTRAN enhancement. Technical report, 3GPP TR 25.892 (2004), V1.1.0, 2004.

[38] 3GPP. LTE physical layer framework for performance verification. Technical report, 3GPP TS G-RAN1#48 R1-070674, February 2007.

[39] G. Calcev, D. Chizhik, B. Goransson, S. Howard, H. Huang, A. Kogianthis, A. F. Molisch, A. L. Moustakas, D. Reed, and Xu Hao. A wideband spatial channel model for system-wide simulations. *IEEE Transactions on Vehicular Technology*, vol. 56, no. 2, March 2007.

[40] ETSI SMG2 Universal Mobile Telecommunications System (UMTS). Selection procedures for the choice of radio transmission technologies of the UMTS. Technical report, ETSI TR 101.112.

11

Conclusions

Shahid Mumtaz and Jonathan Rodriguez

Instituto de Telecomunições, Aveiro, Portugal
e-mail: smumtaz@av.it.pt

Green ICT, or sustainable ICT, is a hot topic and initiative that has recently emerged to address the problematic role of ICT for environmental sustainability. That is, in the past few years, ICT has increased sustainability by decreasing resource intensity but has at the same time encouraged resource consuming lifestyles. Consequently, the aim of green ICT is to make the overall impact of ICT clearly environmentally sustainable and positive.

In the new challenge of energy efficiency in mobile wireless, many challenge have to be addressed, either to the spectrum availability through advanced techniques in wireless optimization. In this book is presented the new challenges in wireless communications for energy efficiency, based on protocols, context, the new short-range based technologies like WiMedia, clustering, new coding technologies, including network coding. Spectral management efficiency, as important role in all presented challenge plays important role in all presented work. However, in the new fields, the evaluation procedures are important to figure the results. We present the methodology for overall performance evaluation, including energy focused performance evaluation. Below is the conclusion for each chapter:

Chapter 2 discussed an accurate closed-form approximation of the SE-EE trade-off, which has not been discussed for multi-cell scenario as very few works have been done so far. Nevertheless, the things we demonstrate in this chapter can also be used for evaluating the impact of using multiple antennas on the EE. Because, EE can mainly be improved through multi-cell system

Shahid Mumtaz and Jonathan Rodriguez (Eds.), Green Communication in 4G Wireless Systems, 259–263.

with diversity such as receive, spatial, transmit diversity. So, existing research of the SE-EE trade-off can be used as a starting point.

Chapter 3 discussed the node discovery is the initial step to initiate communication. In this chapter we have presented our mechanism for context based node discovery, where long range technology (WiMax) supports short range technology (UWB) when discovering a cooperative cluster. To make use of the available beacon IEs of UWB, we have proposed novel IEs representing the energy level and willingness to cooperate of nodes, which further contributes in the cooperative cluster building. We have shown how context list is constructed and how the information is extracted from the context list at MAC layer for further utilization in other operations of the terminal. By means of simulations we showed that our context based discovery mechanism contributes to energy saving in scanning process, and prolongs the MTs overall life time.

Chapter 4 presented an overview of WiMedia MAC protocol. WiMedia offers a distributed UWB MAC with no centralized controller and supports high data rates with the help of a well-defined MAC and physical layer protocols. But the allocation of resources, in a distributed and decentralized network, is still a challenging issue. For the devices to communicate in WiMedia based PANs, we propose three different approaches to analyze the reservation protocol of WiMedia MAC for throughput fairness for both isochronous and asynchronous data traffics. In approach 1, we kept the superframe half reserved for DRP hard and half for PCA using devices. In approach 2 we provide flexibility to DRP based reservation by applying soft reservation and also more MASs for DRP incoming nodes to the system. Based on the limitations of these two approaches, we proposed approach 3 which reserve the MASs using hard, soft and PAC reservation types with standard's bandwidth estimation primitives. The simulation results show that approach 3 provides more fairness to the system compared to the other approaches.

Chapter 5 summarizes work on WiMedia MAC resource allocation protocols. The ECMA standard defined Physical and MAC layers which offer a number of policies and control mechanisms to ensure the quality of service (QoS) provisioning. Although the WiMedia MAC works in a distributed manner and solves the problems encounters while using IEEE 802.15.3, but still there are some minor issues with resource distribution which are highlighted in this chapter. A detail comparison of centralized and distributed resource protocols has been presented followed by details of WiMedia's MAC protocol and resource distribution protocol. A number of unfair distribution (e.g. excessive use of Hard DRP, first come first serve based approach to the resources, etc.) has been elaborated. The existing solution to the unfair distribution of resources from existing research has been provided. Although clustering techniques have been envisaged as a solution for scalability and improving network performance, clustering techniques also presents many possibilities for energy saving applications. The usage of short range communication by multi-hop communication and relaying techniques is currently a hot topic in cooperative systems towards a more energy efficient communication and several solutions have already been proposed in recent years. New hardware capabilities present in smartphones and other portable devices in terms of processing capabilities, wireless interfaces and additional features like GPS have made cooperative techniques to emerge in relevance in the last years. Clustering algorithms have also evolved by using these features in a wide diversity of techniques, looking for energy efficient manners to form stable topologies among mobile devices. Nevertheless, the dynamic nature of mobile ad hoc networks introduces the challenge of adapting the network to fast changes while optimizing the energy consumed, and current clustering algorithms have to face this trade-off.

Chapter 6 discussed the main concepts about clustering techniques and the most known clustering algorithms have been described. Among them, mobility-aware clustering algorithms are explained in detail, since they are normally selected for highly mobile ad hoc networks. Nevertheless, these algorithms still do not solve all problems derived from urban scenarios, whose special characteristics (strong multipath effects, fading, double direction feature of walking pattern, indoor and outdoor environments, etc.) require more considerations at the time of designing clustering algorithms. There is therefore space for future research on this field. Furthermore, multi-user and channel diversity, task scheduling and cognition are open fields where research community has still space for new proposals.

Chapter 7 showed how energy efficiency and reduction of complexity can be achieved by applying Network Coding instead of classical routing in wireless environments. The main scope of this chapter is to be a tutorial, exploring the theoretic framework of Network Coding used for the analysis and the design of efficient solutions in wireless scenarios. The first aim is to introduce hypergraph fundamentals and the main concepts of algebraic Network Coding, adapted to this graph generalization. Then, we provide some key insights on the advantages and drawbacks of random linear Network Coding. After this first general section, the section describing practical Network Coding follows, in which the three main paradigms for RLNC are shown: firstly, the solution which uses coding vectors and generations, simple but with the disadvantage of a significant overhead; secondly, a basic overview of noncoherent Network Coding and Subspace coding theory, which represents the way to remove coding vectors; finally, the recent compression technique that keeps coding vectors and reduce their overhead. In particular, the last method has grown its importance after the discovery that the coding vectors bring passive information on network topology, which can be useful in real scenarios to find failures and bottlenecks. The end of the chapter is equipped with appendices which clarify some theoretic concepts from graph theory and algebra to help the reader to understand the Network Coding results shown along the chapter.

Chapter 8 was based on the assumption that in the near future vertical handovers (VHO) would be a reality. As such, the contents were organized in three main categories: Security, energy efficiency and VHO simulations. Security is a key functionality of any VHO procedure as it would be crucial for the technology to reach its market potential. Toward this goal this chapter presented and described the major efforts done both by the IEEE (802.21a) and IETF (HOKEY WG). On the other hand, the recent concern for green technologies is boosting new research towards energy efficient systems. To this end, the authors in Chapter 8 propose energy saving techniques by exploiting the Media Independent Information Services (IEEE 802.21). It is shown that the use of context awareness is very beneficial in VHO scenarios – to efficiently trigger the mobile's radio interfaces from idle mode and reduce the exhaustive search procedure for new channels. Under these guidelines, mobile devices can save even more energy by being aware of their surroundings, neighbouring mobile devices and access networks. The enabler for this approach would be a context-aware architecture that can push/pull rich context services to the mobile device. The results attained through simulations using an ns-2 based platform show that the proposed approach for energy saving can improve energy consumption, by approximately 30%, on average, with a fairly simple implementation. It is obvious that this energy saving can be even higher if a larger number of different radio interfaces are integrated in the MN.

Chapter 9 presented all the necessary information for modeling and evaluation wireless cellular systems through Link level. Link level evaluates the transmitter and receiver characteristics focusing in the link characterization, namely modulation and channel coding, in terms of energy consumption. The link is analyzed both in transmission and reception point of view. Link level modeling for third and forth generations is presented regarding to the behavior of the link in a computationally efficient way. Link level simulations have a high computational cost and the results are used as input for system level simulation through proper interface, usually by link performance tables in function of SINR values. For energy point of view in link level, we analyze the link adaptation through channel adaptation modulation and channel coding along with automated repeat request protocol. The analysis shown that adaptive link layer transmission taking into account jointly modulation and channel coding rate along with the number of retransmission of the hybrid automated repeat request (H-ARQ) is the most efficient mode in terms of energy and throughput achievement.

Chapter 10 presented all the necessary information for modeling and evaluation wireless cellular systems through System Level Simulator (SLS) for the system perspective in energy consumption evaluation and optimization. It is presented the key aspects on the radio resource management (RRM) in the third and fourth generations that can affect overall system efficiency in terms of energy consumption. Energy efficient resource scheduling are present, including relay based scheduling and how MIMO van be applied to improve the energy consumption. All modeling for the system level evaluation is presented. The minimum system modeling involves three aspects: the users behaviors; the traffics behavior for the application considered and the radio aspects involved in the transmission of the signals from users to destination. To evaluate the performance of SLS a large number of statistics are collected for the computation of the metrics. These performance statistics are generated as outputs from the system level simulations and are used in the performance evaluation of the used scenarios and proposed algorithms. Traffic models and channel models for SLS were also discussed in this chapter. The cellular layout architecture for SLS was fully described. Of particular interest in the realization of system level simulations is the definition of the proper link to system level interface and the definition of the proper set of look up tables used in the mapping of physical layer performance. The procedure followed into the derivation of the Signal to Interference plus Noise Ratio (SINR) and the mapping function used to map the vector of SINRs into a single scalar to be inputted into the look-up tables is detailed. This chapter also describes in detail all steps followed in each frame period by the communication protocol, between the base and the mobile stations, for the transmission of the information associated to the scheduled users.

Author Index

Aguiar, R.L., 9
Alam, M., 33, 53, 69
Albano, M., 33, 53
Bassoli, R., 117
Bastos, J., 139
Huq, K.M.S., 9
Marques, H., 89, 117, 139
Monteiro, V., 177, 203
Mumtaz, S., 1, 9, 69, 177 , 203

Nascimento, A., 203
Politis, C., 177, 203
Radwan, A., 33, 53
Rodriguez, J., 1, 9, 33, 53, 69, 89, 117, 139, 177, 203
Sucasas, V., 89
Tafazolli, R., 89, 117, 139
Verikoukis, C., 69

Subject Index

4G, 9
802.21, 139
ad hoc networks, 89
beacons, 69
beyond 3G, 177, 203
clustering, 89
cross-layer, 203
decentralized MAC, 69
distributed reservation protocol, 69
energy efficiency, 9, 33, 89, 139
green, 9, 89
handover, 139
heterogeneous networks, 139
link-level, 177
LTE, 177
medium access control, 53

MIH, 139
multi-RAT, 177, 203
network coding, 117
node discovery, 33
piconet, 33
practical network coding, 117
quality of services, 53
resource allocation, 69
simulation, 177, 203
spectral-efficiency, 9
subspace nodes, 117
system-level, 203
ultra wideband, 33, 53
WiMedia, 53
wireless networks, 117, 139
Wireless Personal Area Networks, 33

About the Editors

Shahid Mumtaz received his Master and Ph.D. degree in Electrical & Electronic Engineering from Blekinge Institute of Technology Karlskrona, Sweden and University of Aveiro, Portugal in 2006 and in 2011. Since then he has joined Instituto de Telecomunicações, working as a senior researcher in 4Tell Group. He was with Ericsson Research Labs in 2005, working as a research engineer. His research interests include green communications, cognitive radio, cooperative networking, radio resource management, cross-layer design, heterogeneous networks, M2M and D2D communication, and baseband digital signal processing. He has more than 30 publications in international conferences, journal papers and book chapters. He is a guest editor for one special issue in *IEEE Wireless Communications Magazine and Communication Magazine*.

Jonathan Rodriguez received his Masters degree in Electronic and Electrical Engineering and Ph.D from the University of Surrey (UK), in 1998 and 2004 respectively. In 2002, he became a Research Fellow at the Centre for Communication Systems Research and was responsible for coordinating Surrey involvement in European research projects under framework 5 and 6. Since 2005, he is a Senior Researcher at the Instituto de Telecomunicações (Portugal), and founded the 4TELL Wireless Communication Research Group in 2008. He is currently the project coordinator for the Seventh Framework C2POWER project, and technical manager for COGEU. He is author of more than 170 scientific publications, served as general chair for several prestigious conferences and workshops, and has carried out consultancy for major manufacturers participating in DVB-T/H and HS-UPA standardization. His research interests include green communications, network coding, cognitive radio, cooperative networking, radio resource management, and cross-layer design.